T0260797

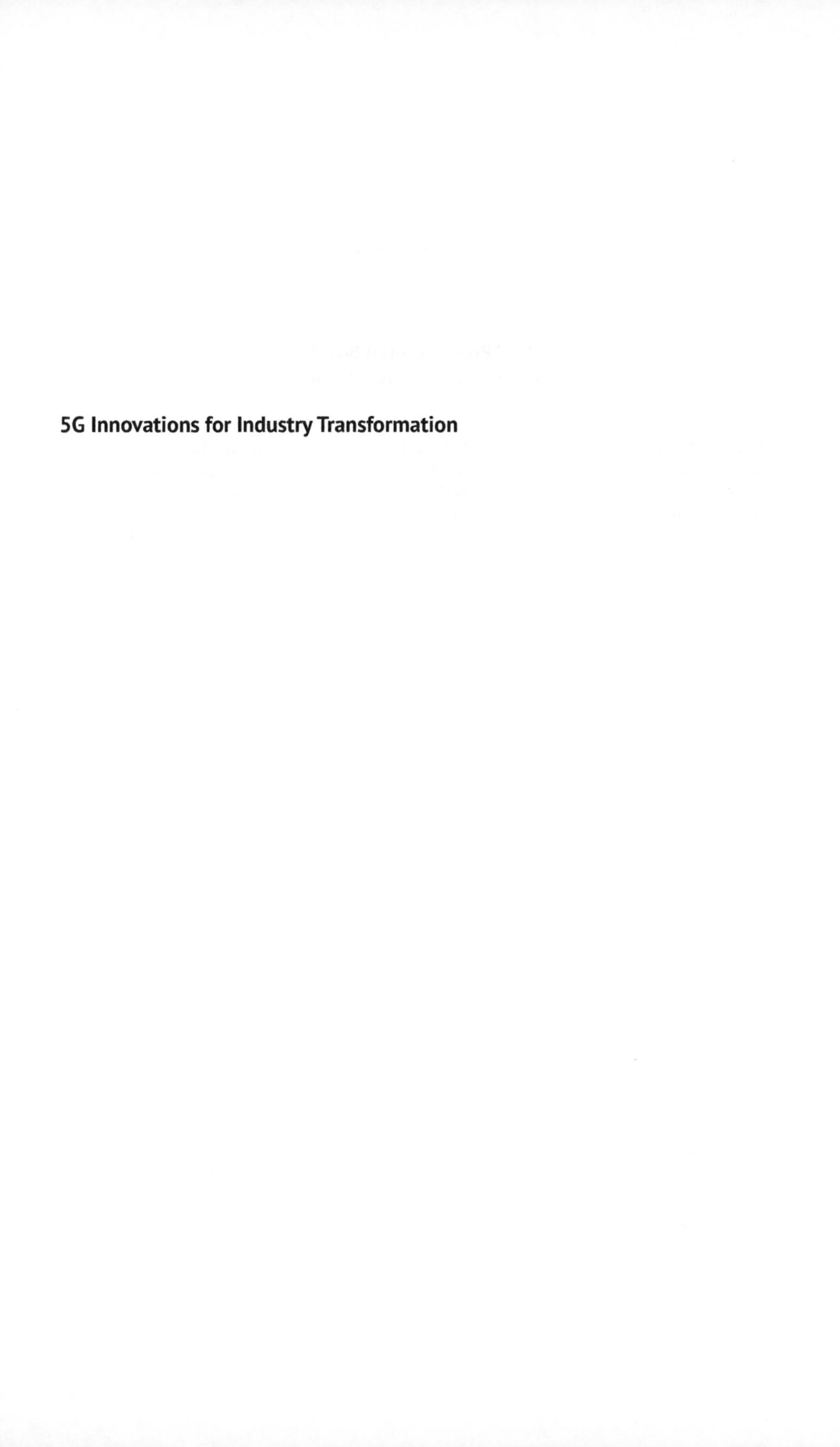

5G Innovations for Industry Transformation

IEEE Press
445 Hoes Lane
Piscataway, NJ 08854

5G Innovations for Industry Transformation

Data-Driven Use Cases

Jari Collin
Adjunct Professor at Aalto University and CTO at Telia Finland

Jarkko Pellikka
Program Director, Mobile Networks, Nokia

Jyrki T.J. Penttinen
Technical Director

Published by John Wiley & Sons, Inc., Hoboken, New Jersey.
Published simultaneously in Canada.

For general information on our other products and services or for technical support, please contact our Customer Care Department within the United States at (800) 762-2974, outside the United States at (317) 572-3993, or fax (317) 572-4002.

Wiley also publishes its books in a variety of electronic formats. Some content that appears in print may not be available in electronic formats. For more information about Wiley products, visit our website at www.wiley.com.

Library of Congress Cataloging-in-Publication Data Applied for:

Hardback ISBN: 9781394181483

Cover Design: Wiley
Cover Image: © dem10/Getty Images

Set in 9.5/12.5pt STIXTwoText by Straive, Pondicherry, India

Contents

About the Authors

Prof. Jari Collin is an adjunct professor of Enterprise Information Systems and Service Networks at Aalto University, Finland. His areas of specialization include industrial internet, digital services, and management of demand-supply networks. He obtained his M.Sc. degree in Industrial Management from Tampere University of Technology in 1996 and his D.Sc. degree in Industrial Management from Helsinki University of Technology (currently part of Aalto University) in 2003. Professor Collin has published a number of journal articles and conference papers as well as two books on industry digitalization. His current research is centered around the applications of industrial internet and 5G/6G in boosting digital transformation with new data-driven services.

Dr. Collin has more than 25 years of experience in the telecom and ICT industry. Currently, he works at Telia Finland as Chief Technology Officer (CTO) and heads the company's Infrastructure unit.

Dr. Jarkko Pellikka is the leader of Nokia's major R&D initiatives on Industrial 5G and edge computing. He has extensive experience in technology management, innovation ecosystems, and commercialization of innovation. Jarkko Pellikka has worked for several years as senior leader in global multinational companies as well as with startups and SMEs being responsible for leading and developing numerous strategic transformations in multiple industries. His experiences and thoughts have been published in several scientific international journals and books.

Connect with Jarkko on LinkedIn

Dr. Jyrki T.J. Penttinen has worked for mobile network operators and device manufacturers, security and roaming providers, and membership organizations in Finland, Spain, Mexico, and the United States since 1994. He is experienced in research and operational activities covering mobile network design and performance aspects, standardization, services, and product development. Dr. Penttinen is also a published author and instructor of telecommunication technologies.

Foreword

The Nordic countries are consistently ranked among the most digitally advanced in the world, with high rates of internet penetration, smartphone adoption, and digital skills. In the early days of mobile telephony, Nordic nations were among the first to deploy cellular networks and collaborate on the development of common standards, such as the analog Nordic mobile telephone (NMT) system, used from the 1980s to the early 2000s. In the 1980s, they were also involved in the development of the Global System for Mobile communications (GSMs) standard for second-generation (2G) digital cellular networks, and they continued to play leading roles in the development of more recent cellular standards, such as 3G, 4G, and 5G.

Nordic countries have been able to play a leading role in setting cellular radio technology standards because they have a strong telecommunications industry and a commitment to innovation. They have also been willing to collaborate with other countries to develop standards that are widely adopted. Today, thanks to those principles of innovation and collaboration, 5G cellular network technology is revolutionizing the way people around the world live and work. With its significantly faster speeds, lower latency, and greater bandwidth, 5G is enabling a wide range of new and innovative applications in a variety of industries.

In the new digital economy, connectivity has become a strategic priority that business leaders need to pay attention to. The fifth generation of cellular network technology allows enterprises to collect and analyze previously unthinkable amounts of data, which can be used to improve decision-making, increase sustainability, and, ultimately, produce better products and services. By giving customers the ability to interact with businesses anytime, anywhere, 5G connectivity leads to a better customer experience. Connectivity helps businesses to collaborate with partners and suppliers more effectively, reach new markets, and develop new products and services, and it helps to improve efficiency by automating processes and enabling collaboration between employes.

In the mining industry, for example, readers of this book will discover that 5G is expected to radically improve opportunities to wirelessly automate operations both underground and on the surface. Compared to 4G, 5G provides superior uplink throughput, which is vital when operating underground. While this uplink throughput can be matched by a well-customized Wi-Fi network, 5G also offers superior connection reliability. Increased use of 5G in mining promises to increase safety in what can be a hazardous work environment.

The forestry industry is using 5G to connect and control robots and other machinery in real-time, enabling more efficient and productive operations. In one promising example, a robot dog called Frans participated in an experiment at a pulp mill where it performed tasks such as measuring the temperatures of various components with its thermal imager. In the near future, robots like Frans, leveraging 5G and edge computing, could perform tasks that are dangerous, repetitive, or require extreme precision.

Elsewhere, elevators might not be the first place you would look for innovation – but 5G standalone technology is being used to enhance the capabilities of a concept elevator that can be installed early on in a construction project and then used to optimize material flows during the construction process. Data from a 360° camera and sensors attached to machines, tools, and workers are collected and wirelessly transferred using 5G standalone technology and a dedicated network slice, ensuring isolated end-to-end connection capacity. This data are then analyzed using machine learning and artificial intelligence to optimize the flow of material and workers at the construction site.

One of the biggest challenges facing our own industry, telecoms, is ensuring that the energy consumed by our networks does not increase at the same rate as data flows. So it is helpful that 5G introduces many new energy-saving features. In the radio access network, which is responsible for around 80% of all electricity consumed by a mobile network, 5G offers higher spectral efficiency and makes it easier to power down equipment when it is not in use.

In the oil and gas industry, energy providers are using drones for video surveillance of pipelines, plants, and infrastructure, to increase both safety and efficiency. Real-time streaming of high-definition video from drones, combined with analytics for detection, can help energy providers identify risks and defects, such as a leaking pipe. Whereas a private LTE network could support full HD video streams, a 5G network is required to support simultaneous 4K video and data sensor streams. Other 5G benefits include enhanced data security and reliability, and low latency to support increased density of remote applications and video analytics.

As these use cases demonstrate, industrial 5G is not just solving business challenges, but also contributing to sustainability. The real-time monitoring and control of industrial processes enabled by 5G is helping to improve energy efficiency and lower greenhouse gas emissions, thereby reducing the overall environmental impact of industrial operations. 5G-enabled predictive maintenance is helping to reduce downtime by detecting potential issues before they develop into major problems, contributing to improved equipment performance and reduced waste of resources. By supporting the use of real-time data and information in decision-making and safety processes, 5G is also improving safety in industrial settings. In addition, 5G enables increased automation of industrial processes, helping to reduce the risk of human error.

Finally, advances in the development of 5G, the Internet of Things, and cloud technologies have accelerated the transition from traditional logistics and supply chain management towards Industry 4.0, in which everything across all industries is connected via a standardized, secure, and reliable wireless communications system. Real-time tracking and monitoring of products and materials, enabled by 5G, is improving supply chain management, optimizing transportation, and reducing waste.

The potential industrial innovations enabled by 5G are limited only by our imagination. In this book, Jari Collin, Jarkko Pellikka and Jyrki T.J. Penttinen have done a fantastic job of providing a comprehensive overview of the impact that the latest 5G technologies are having on a variety of industries. We trust that it will inspire the next generation of big thinkers in telecom and beyond to stretch the limits of 5G, and future cellular technologies, even further.

Allison Kirkby	**Pekka Lundmark**
President and CEO	President and CEO
Telia Company	Nokia

Preface

As Alexander Graham Bell said, "*Great discoveries and improvements invariably involve the cooperation of many minds.*" This has been our main guiding principle since we started in 2021 this common journey with the leading global companies of five selected industry sectors. Based on contemporary cross-industry experiences, our aim was to identify and empirically test how 5G innovations can boost the ongoing digital transformation. We commonly agreed on a pragmatic approach to first understand specific needs, requirements, and opportunities in each selected industry vertical and then share best practices between the industries. Focus was to help companies to better utilize real-time data in achieving their high-standard business objectives on productivity, safety, and sustainability. We conducted multiple case studies to design and test new data-driven use cases together with the enterprises from mining, elevator, forest, telecommunications, and oil refining industries. We hope that our efforts with the leading companies provide useful insights on how to leverage 5G in industry transformation.

The case studies of the book are based on a co-creation research project as part of Nokia's (Veturi) initiative entitled "Unlocking Industrial 5G" that explores how to leverage 5G technology in digital transformation across industries. We attempted to provide a holistic perspective on the topic by describing the main barriers to creating data-driven services in industrial context. In addition, we wanted to seek practical ways to overcome these challenges by also introducing a set of prerequisites and recommendations based on the empirical evidence on the successful 5G implementation in the ecosystem context. As the case studies indicate, well-planned and orchestrated ecosystem collaboration is the key to effectively combining the state-of-the-art capabilities from the different organizations to promote actual value capturing. The current and future challenges to digital transformation in the industrial context are very complex and challenges the conventional ways to lead collaboration needed to unlock industrial 5G. Therefore, we attempt to provide practical guidelines and ideas on how to benefit from 5G technology throughout industry-wide digital transformation. Examples of such opportunities are automated quality and safety control, predictive maintenance with online asset management, real-time situational awareness, and mobile working vehicles in an industry area. Sharing learnings and best practices between different industry sectors forms an essential part of the book and serves especially decision-makers across industries as a practical "travel guide" to utilize the multiple new capabilities of 5G technology for digital transformation.

5G provides industrial companies with an open and trusted digital platform to accelerate operational innovations and new business models in the whole industry ecosystem. After decades of relatively incremental evolution of the mobile generations, the new 5G systems provide operators and cooperating parties with totally new means that benefit the whole ecosystem by offering disruptive services and solutions via, e.g. evolved means to expose the network functions and capacity to third-party service providers that, in turn, can create totally new businesses and revenue sharing models. Also, the new features of 5G are designed to meet the highly varying needs and requirements of verticals. With such fast pace of the technologies, there is not too much information available on the solutions and how ecosystem can apply those in practice. Various new stakeholders have an excellent opportunity to join ecosystem collaboration and cocreate new solutions based on the enablers created by 3GPP and other standards developing organizations.

Finally, we know that this book provides only a modest increase to the existing body of knowledge and many questions have not been answered yet. Therefore, more efforts, studies, and ecosystem collaboration are needed to unlock the full potential of industrial 5G.

September 3, 2023

Jari Collin, Jarkko Pellikka, and Jyrki T.J. Penttinen

Acknowledgments

First and foremost, we want to express our deepest gratitude to the case study companies and their high-skilled representatives with whom we designed and implemented the data-driven use cases. These forerunner companies are globally known for their innovative ways of utilizing digitalization and real-time data in their industry ecosystems to improve customer value, productivity, and/or long-term sustainability. The companies together, with Nokia, hosted eye-opening and educational cross-industry workshops to share best practices.

Without the empiric use cases and shared industry learnings, we would not have sufficient empirical research data on how to leverage 5G technology in digital transformation across different industry ecosystems. Here, the master's thesis workers of Aalto University played a central role in collecting and analyzing the research data in these company case studies. Their contribution has been essential for our book's case study part.

The following key people made an important contribution to the case studies, and they all deserve big thanks:

- Sandvik is a digital forerunner in the mining industry, and its use cases were striving for autonomous connected operations underground. The benefits of 5G technology compared to existing Wi-Fi-based applications were identified and quantified in the case study. The core team consisted of **Miika Kaski**, **Teemu Härkönen**, **Ville Svensberg**, **Jyrki Salmi**, and **Jagdeesh Rajani** (thesis worker).
- Metsä Group represents the Finnish forestry industry, and the case study focused on opportunities and barriers of 5G to improve productivity in modern bioproduct mill operations. The team included **Jani Salonen**, **Janne Pekola**, **Jukka Mokkila**, and **Perttu Laiho** (thesis worker).
- Kone represents a global elevator technology industry. The study concentrated on buildings construction phase, during which smart elevators can provide the ecosystem with a common digital platform to optimize logistics on construction sites. This innovative team consisted of **Janne Öfversten**, **Mika Kemppainen**, **Tommi Loukas**, and **Ella Koivula** (thesis worker).

- Telia Finland represents telecom industry with the case study on how to improve energy efficiency for climate. The comparison of the energy efficiency between 5G versus 4G networks was studied by using network data analysis. The team was composed of **Janne Koistinen**, **Eija Pitkänen**, **Timo Saxen**, and **Roope Lahti** (thesis worker).
- Neste represents modern oil refinery industry, and the case study focused on improving operations with 5G-enabled drones at a refinery area. The team consisted of **Visa Oksa**, **Janne Anttila**, **Jari Manninen**, and **Maarten van der Laars** (thesis worker).

We also want to thank the book's reviewers and Aalto University's Prof. Raimo Kantola, Prof. Robin Gustafsson, Dr. Jose Costa Requena, and Dr. Kari Hiekkanen who supported us in forming a professional research agenda and executing the multiple case study research – likewise, many other colleagues at Aalto and Nokia who helped us in the journey. In addition, we are very grateful to Business Finland for funding this co-creation research project with Aalto University and Nokia Veturi program.

The whole Wiley editing team deserves a special acknowledgment for their professional support throughout the book editing process.

Finally, we like to present our highest appreciation to Allison Kirkby (Telia Company, President and CEO) and Pekka Lundmark (Nokia, President and CEO) for paving the way in industrial 5G and writing the joint foreword for our book!

September 7, 2023

Jari Collin, Jarkko Pellikka, and Jyrki T.J. Penttinen

Part I

New Data-Driven Business Opportunities with Industrial 5G

1

Digital Disruption of Industries

1.1 Introduction

The digitalization of products and services is increasingly disrupting competition and the existing industry borders [1, 2]. Traditional physical product-based business models are being challenged by data-driven digital ecosystems that pursue new ways to increase end-customer value. Modern digital platforms and the use of real-time data enable a substantial leap in value – as already witnessed in numerous Internet of Things (IoT) applications for consumers. New IoT-enabled services extend a value offering from delivering products to using them in the most optimal way. Increased customer value is created with online services at the very moment when a consumer uses a product in that unique situation and environment [3]. As the father of modern management, Peter Drucker [4], aptly pointed out decades ago, "what the customer considers value is not a product itself but utility, that is, what a product does for him/her."

In the enterprise markets, a lack of common business rules for sharing data between enterprises has hindered a similar development in Industry IoT (IIoT) applications, although business opportunities with data-driven online services are evident. In addition, data security risks and business continuity requirements in mission-critical processes can easily become barriers to the extensive use of cross-company data. However, successful trials and pilots among forerunner enterprises and ecosystems exist with promising results. The industrialization of these lessons requires a common, trusted connectivity and computing platform for managing shared data in real time. The fifth-generation mobile technology (5G) enables such an open and trusted platform for industrial ecosystems to operate safely

5G Innovations for Industry Transformation: Data-Driven Use Cases, First Edition.
Jari Collin, Jarkko Pellikka, and Jyrki T.J. Penttinen.

Published 2024 by John Wiley & Sons, Inc.

together globally. In addition, a close ecosystem-wide collaboration and willingness to learn from other industry ecosystems are important to make digital transformation happen.

There are industry ecosystems where digitalization is already mature enough to gain industry-wide tangible benefits. Telecom, banking and insurance, and media businesses are examples of such pioneering industries that have already revolutionized business models in these sectors [5]. The utilization of digitalization is also essential for the competitiveness of companies in these sectors, as the sectors are also subject to international competition and their competitors increasingly benefit from digitalization. However, the productivity benefits of digitalization vary dramatically by sector. The appearance of digitalization in everyday business can look very different in various business and public sectors. International studies confirm the view that these differences are due to the ability of different sectors to utilize digitalization: banking operations are easier to digitalize than a construction site [2]. A report published by the Organization for Economic Co-operation and Development (OECD) indicates that the information and financing sectors are those that are furthest ahead in overall digitalization [6]. The differences between companies in the same sector can also be extensive [2].

This book is about industry (digital) transformation and data-driven online services boosted by 5G technology. It is written based on contemporary research findings and practical lessons on how to leverage 5G in industry transformation. The aim is to describe proven data-driven use cases that utilize industrial 5G in driving customer value, productivity, and/or sustainability in selected industry ecosystems. The research includes five different industry sectors: mining, forest, lift, telecom, and oil and gas industries. The viewpoint is an industrial enterprise that seeks new business opportunities with industrial 5G. Sharing lessons and best practices between different industry sectors is an essential part of the book.

1.2 Industrial 5G Boosts Digital Transformation

1.2.1 Industry 4.0 – The Ongoing Industrial Revolution

We are witnessing the Fourth Industrial Revolution (referred to as Industry 4.0) that offers huge opportunities for multiple industries to improve radically the productivity and sustainability of their global operations. The concept of Industry 4.0 has its roots back to 2011, when it was publicly introduced for the first time at Hannover Messe in Germany. In 2013, Professor Henning Kagermann and his highly rated research team provided the German government with *Recommendations for implementing the strategic initiative Industry 4.0* to secure the future competitiveness of the German manufacturing industry [7].

Originally, the concept emerged around smart manufacturing with new real-time capabilities for vertical integration, horizontal integration, and end-to-end digital engineering. The ongoing industrial revolution is based on cyber-physical systems (CPS) and industry digitalization. It is powered by both established and emerging technologies, including, for instance, IoT, artificial intelligence (AI), advanced data analytics, robotic process automation, robotics, cloud computing, virtual and augmented reality (VR/AR), 3D printing, and drones [8].

Since those days, the concept of Industry 4.0 has been widely studied from numerous perspectives. As our viewpoint in the book represents a forerunner enterprise that plays a significant role in its industry ecosystem to seek new business opportunities and benefits, there are three interesting perspectives to Industry 4.0 implementation: (1) business potential, (2) technology elements, and (3) implementation design principles [9, 10]. The business perspective includes key changes that Industry 4.0 brings to business to improve the value creation of a company in its value chain by adopting digital technologies [11]. At a high level, these changes can increase efficiency and provide a strategic edge over competitors by changing business models, improving customer service, and helping adjust to labor market changes [12–14]. The key elements of Industry 4.0 cover a wide range of technologies. CPS, IoT, Internet of Services (IoS), and smart factories have all been identified as the higher-level key elements that cover the fundamental technologies [15]. The implementation of Industry 4.0 is guided by a set of key principles that determine the basic idea and mechanisms. Technical assistance, decentralized decisions, interconnection, and information transparency are the main items guiding the implementation [16]. These three perspectives are not independent but closely related to each other [9].

Industry 4.0 is strongly associated with the new generation of industrial automation resulting from digitalization and analytics. At the heart of the thinking is the product's manufacturing technology and factory production, as well as its connection to the Internet. New cyber-physical solutions connect people, products, and services. The businesses of the future will be part of a global network and form autonomous entities where intelligent machines, production processes, and warehouses will work together in real time to improve product lifecycle and supply chain management.

Industry 4.0 has also been popularized under different names in different parts of the world, e.g. the term IIoT, also referred to as Industrial Internet, is often used in the United States. In this context, IIoT as a term refers to IoT technology used in industrial settings. However, it is not limited to any specific industry sectors, e.g. manufacturing, but covers all segments having industrial operations – from dairies, slaughterhouses, and bakeries to heavy industries, such as steel mills, paper mills, machine shops, mines, and power plants [3]. The IIoT consortium, established in 2014 as a global not-for-profit partnership of industry, government, and academia, defines the term as follows: "Industrial Internet is an internet of things, machines, computers, and people enabling intelligent industrial operations using advanced data analytics for transformational business outcomes" [14, 17]. An Industrial Internet solution contains four key characteristics: real-time data processing, transaction predictability, mobility of operations, and increased automation [3].

1.2.2 Digital Transformation

Digital transformation refers to the economic and societal effects of digitization and digitalization [18]. Digitization is the conversion of analog data and processes into a machine-readable format. Digitalization is the use of digital technologies and data as well as their interconnection, which results in new or changes to existing activities. Digital transformation is characterized by a fusion of advanced technologies and the integration of physical and digital systems, the predominance of innovative business models and new processes, and the creation of smart products and services [19]. Digital transformation is based on using digital technologies to change the practices, processes, organization, and value creation of an organization, sector, or industry [20].

MIT University in the United States has been studying digital business for over 20 years. Together with Capgemini, researchers have conducted empirical research into how pioneering companies have successfully led a digital transformation program in their organizations [21]. Numerous use cases and best practices exist to explain why organizations need to embrace digital tools to stay competitive, and how to become a "digital master." Despite digital technologies, digital transformation is much more than an IT upgrade as the focus is on changing the fundamental parts of an organization such as strategy, value creation, and organization to enable new forms of doing business and achieving competitive advantage [3, 5, 22].

Managing digital transformation does not fundamentally differ from a company-wide change program, and the same leadership and management principles are valid. Top management involvement in defining strategic direction and commitment in execution is essential. The keys to success consist of the following four steps: (1) identify opportunities and renew strategy; (2) brainstorm, make proof-of-concept, pilot, and analyze the results; (3) refine strategy and lead change; and (4) avoid pitfalls and create best practices [3].

1.2.3 Industrial 5G

Connectivity is one key enabler that allows these digital technologies to realize their full potential. The new mobile technology 5G enables significant improvements in connectivity but also provides an open, trusted data platform for industrial ecosystems to operate safely together. Historically, industrial revolutions have been characterized by the transformation of physical infrastructure networks [7]. As a means of transmitting data from producers to consumers, 5G definitely has a role to play in ensuring trust between stakeholders. An effective use of 5G networks is a promising opportunity to enhance trust further that can take various forms for the various parties involved in a digital ecosystem as follows:

- Individuals or organizations, which are the sources of the data, are concerned whether organizations that process data use the data as authorized.
- An organization that processes the data are concerned about data provenance.
- Individuals care that data are used only for purposes that have been clearly stated.
- Organizations that use data output must rely on the output being correct and unbiased.

Therefore, trust-intensive data management is expected to be applied from the business perspective in the context of 5G, also creating new opportunities for multiple organizations, including infrastructure providers, connectivity service providers, data service providers, and integrators.

The Fourth Industrial Revolution's potential can be realized through the wide-scale planning and deployment of 5G communication networks in order to realize the following benefits of 5G capabilities [23–26]: high-speed broadband, ultrareliable low-latency communication (URLLC), massive machine-type communications, high reliability/availability, and efficient energy usage. Today, many companies are deepening the integration between industrial automation systems and enterprise applications to improve efficiency further throughout the value chain. Information and communications technology (ICT)/operational technology (OT) convergence, connected factory, connected enterprise, IIoT, Industry 4.0, and smart factories are all concepts that are part of the ongoing evolution of industrial automation. Industrial automation systems have traditionally relied on hierarchical, siloed communications between control and field devices using industrial protocols. Development of proprietary technologies with internet protocol (IP)-based standards can make industrial automation systems and related devices more interoperable and more consumable for multiple industrial usage [27, 28]. Increased interoperability and communications between devices, systems, services, and people in combination with technologies such as advanced sensors, smart devices, and wireless technologies with machine learning (ML)/AI-enabled capabilities improve performance, flexibility, and responsiveness throughout the value chain. An increasing number of sensors, tags, miniaturized computers, transmitters, and network technology can, for example, enable unfinished products to send data to machines about what is needed for them to be completed.

In addition, 5G enables new network capabilities that are essential, especially in the industry domain, such as network slicing and edge computing. All these new functionalities are also increasing industry interest in private 5G networks for delivering cellular connectivity for many private network industrial use cases. From this perspective, the term "Industrial 5G" can be defined including the key elements of IIoT as follows: "An entity that combines large number of networked sensors, assets and objects driven by data that are connected using 5G technologies, related wireless communication systems, and edge computing platforms to enable real-time, very reliable, low-latency and high-bandwidth data transmission and communication, associated generic information technologies and optional cloud or edge computing platforms" [also see Refs. 2, 3]. However, the investments in Industry 4.0 solutions and IoT applications have not yet delivered on their promise of increased productivity and competitiveness. A great deal of value is still waiting to be captured in the coming years. It has the potential to transform industries by ensuring the 5G radio network coverage, capacity, and quality needed for new business models and industry applications. According to [2], 5G will contribute to industrial advances in three significant ways by (1) enabling faster and effective inspections through predictive intelligence, (2) improving workplace and worker safety, and (3) enhancing operational effectiveness. 5G also has the potential to impact industry by managing its carbon footprint and improving energy efficiency.

5G drives digitalization of industrial sites across different industrial segments. It has the potential to lift global economies by sparking a huge increase in productivity in a sustainable way. 5G dramatically reduces the energy needed to transfer bits of data. Significant economic and social value can be gained from the widespread deployment of 5G networks. Technological applications, enabled by a set of key functional features, will both facilitate industrial advances, improving productivity and profitability, and enhance city and citizen experiences. To accelerate the adoption of 5G, new ecosystem-based collaboration, business models, and agile practices among stakeholders will be needed, along with clear methodologies to estimate the social value creation to enhance the business case of 5G. 5G growth opportunity comes with new customer segments for industrial 5G and opportunities for new business models and new 5G-enabled digital services. In fact, the telecom and network business will change with 5G. New products and business models are needed to transform the macro-network business, especially into industry-specific business models with tailored offerings. According to the report published by [29], the global industrial 5G market generated US$12.47 billion in 2020 and is estimated to earn US$140.88 billion by 2030, indicating a cumulative annual growth rate (CAGR) of 27.5% from 2020 to 2030. For example, increase in demand for high-latency and low-latency networks among various industries, rise in machine-to-machine (M2M) connections across various industries, and demand for next-generation telecommunication network service among enterprises all drive the growth of the global industrial 5G market. However, the high implementation cost of 5G solutions hinders market growth. On the other hand, the development of smart infrastructure such as 5G-enabled facilities and adoption of IoT-based 5G infrastructure across various enterprises present new opportunities in the coming years.

The adoption of wireless solutions in industrial environments is often a gradual process, and an initial deployment typically comprises clusters of wireless devices connected to an existing wired network. Although wired networking solutions are still predominantly used for industrial communications between sensors, controllers, and systems, wireless solutions will be more often used, for example, in areas that are challenging to reach and/or due to safety reasons. Proprietary radio solutions have traditionally been used to support these use cases and are still used in many applications today. However, 5G across industrial domains will create several opportunities to respond to the real needs for bandwidth, latency, or capacity. For example, [11] estimates that annual shipments of wireless devices for industrial automation applications, including both network and automation equipment, reached 4.6 million units worldwide in 2018, accounting for approximately 6% of all new connected nodes [30]. Growing at a compound annual growth rate (CAGR) of 16.3%, annual shipments are expected to reach 9.9 million in 2023. The installed base of wireless devices in industrial automation applications is forecasted to grow from an estimated 21.3 million connections at the end of 2018 to 50.3 million connected devices by 2023.

Therefore, we believe that a significant economic and social value can be gained from the widespread deployment of 5G networks, but it needs decision-makers to understand what the key drivers are, and how they impact on the value creation and digital transformation across industries and companies. From the management point of view, 5G can deliver a high-speed, reliable, and secure broadband experience and will be a major technology that accelerates industry digitization and the massive rollout of intelligent IoT, and thus enables improvements to productivity and sustainability through widespread adoption of critical communications services [31].

The key benefits and the fundamentals to drive 5G-enabled digital transition across industries will be briefly described next.

1.2.4 New Technology Capabilities from 5G

Adaptive tailoring for network hardware, software, platforms, and applications is needed for industrial use cases. For example, enhanced mobile broadband, URLLC, security, massive machine-type communications, and power efficiency are the key 5G-related functionalities that drive the wider utilization across the industries [32]. New use cases and digital services enabled by 5G's ultra reliable low latency communication (URLLC) involving sensor networks and IoT will open new opportunities for communication service providers (CSP) as well as other organizations such as web scalers. For example, 5G-enabled edge clouds bring control and processing very close to industrial machines in order to help reduce cost, reduce latency, and increase speed, especially for mission-critical processes. Another key capability is 5G slicing that enables customized quality of service from device to application by creating virtual network instances that guarantee performance requirements for specific industrial applications as well as isolate traffic and resources needed in the different industrial domains. In practice, 5G can be integrated in time-sensitive networking (TSN) and provide synchronized wireless communication; with URLLC, it can deliver a maximum of 10 ms latency for critical industrial applications [33].

Organizations must prepare their networks for the scale and flexibility that are required to provide highly cost-effective solutions that support exponential increases in network demand, a wide variety of devices and applications, higher data rates, improved data privacy and security, lower latency, and greater power efficiencies [3]. For example, network slicing capability as part of 5G technology is a key enabler for new 5G-based services, enabling multiple stakeholders to use digital infrastructure to establish "slices" of their networks through virtualization technology. These slices can be planned and modified to respond to different requirements (on demand) across industry sectors. This will help transform various industries and create new business models for the telecom industry. It will also help to handle the huge variety of 5G services with different requirements, thus raising the networks' productivity and opening up a new opportunity to innovate their business models to monetize the opportunities [34]. In addition, new functionalities of 5G help industries to respond to their objectives, e.g. through the following: (1) massive multiple-input and multiple-output (mMIMO) and beamforming that enable an increase in capacity of the networks; (2) extended frequencies that provide a better mix of 5G coverage and capacity; (3) integrated access and backhaul (IAB) enables, e.g. lower cost and faster high-density deployments; (4) Sidelink enables device peer-to-peer communication high-density deployments; and (5) URRC enables critical and time-sensitive applications enabling, e.g. IoT devices to deliver significant amounts of data with a latency of 10–30 ms. In addition, new radio (5G) (NR) positioning will provide increasing levels of positioning accuracy below 1.0 m that can be used for different use cases targeting improved safety and productivity, e.g. through asset tracking in the industrial environment.

1.2.5 Business Model Disruption

The current traditional model of wireless communication cannot provide all the potential benefits for industrial sectors with a myriad of IoT devices to be connected. To respond to the business requirements of these industrial sectors, the on-demand approach and related tailored private wireless network model have received increasing attention that challenges the current traditional business models [35, 36]. This development will change traditional business models and ecosystem roles as well as create the basis for a new mobile network operator model, i.e. local and/or micro-operators [36]. These can target specific customers in different industrial verticals with closed 5G networks, serve CSP's customers in high-demand areas on behalf of the CSPs as a neutral host with open 5G networks, or mix different types of customers and offerings through various ecosystemic business models [37]. This described digital transformation across industries will drive the creation of a new generation of network infrastructure that will be built with 5G to complement the current macro-level networks. It has been estimated that this transformation enabled by 5G will foster socioeconomic growth in the fourth industrial revolution with an estimated US$13,200 billion of global economic value reached by 2035 and generate 22.3 million jobs in the 5G global value chain alone [1]. This transformation has accelerated digitalization for local service delivery as well as boosted local and regional businesses into new growth areas, e.g. through new data-driven digital services. Micro-operators can provide locally hosted connectivity with the customized digital services [33, 35, 36] if the viable business model can be defined and implemented. To create and capture value for customers in the industrial verticals, the business model must support real-time interoperability with operations, employees, customers, partners, and suppliers.

5G will also transform the wireless communication ecosystem by introducing, e.g. location-specific private wireless networks that can be operated by different stakeholders. Also here, the objective of the ecosystem strategies is to enable and engage value-adding ecosystem partners to develop the new business and related markets together [7, 27]. These new business models enable companies to fully control all data traffic and applications [32]. Private networks are built on demand for specific industry use cases. In practice, they can help companies to plan and build the industry-specific systems that integrate, for example, manufacturing machines, sensors, field devices, and professionals. All of these will not only respond to the key requirements listed above but also help realize the benefits of wireless broadband, ubiquitous coverage, and mobility to drive new innovations and business models and overall speed up the adoption of the IoT. Industry 4.0 has already brought about various data platforms and related services to the industrial context [35]. Fifth-generation mobile technology and its promises of improvements in supporting critical and massive M2M communications may mean major productivity and sustainability-related enhancements. In vertical industrial contexts, the role of local and private 5G networks has emerged as an increasingly important topic [35]. Business models describe the rationale of how organizations create, deliver, and capture value with data. The traditional business models are changing since many anticipate that some of the most interesting and important applications for 5G will be in the vertical industrial sectors and other private-network applications [32].

1.3 Toward New Business Models

5G technologies are expected to transform future wireless networks in five areas: (1) densification and extreme capacity through millimeter-wave small cells in the access network; (2) localization through the distribution of radio and core functions, content, and services on edge networks to pool gains, and to achieve low latency, high reliability, security, and privacy; (3) decomposition of network functions utilizing interconnected distributed data centers and cloud infrastructure to increase flexibility and scalability; (4) softwarization of the network with advances in analytics and machine learning to enable a high level of automatization in management and orchestration; and (5) network virtualization, particularly network slicing, utilizing the above capabilities to enable various new as-a-service business models [38, 39]. For these kinds of highly localized and heterogeneous environments where security, privacy, and vertical-specific and user-specific requirements play an important role, private local 5G networks [40] can be one alternative.

In this chapter, the term "business model" refers to the logic behind the selected approach in order to create and capture the value of the selected customer segments through a commercialization process [41]. According to [24], elements of the business model are:

1) Customer value proposition (CVP): the methods used to help customers solve an essential business-related challenge or to deliver value to their business.
2) Profit formula [42]: the plan describing how the enterprise creates value for itself while providing value for the customer. It may include information on the revenue/monetizing model, cost structure, margin model, and resource velocity.
3) Key resources: the assets, such as people, technology, products, and equipment, required to deliver the value proposition to the targeted customers.

In addition, in order to create and capture value for the customers in the industrial verticals, the business model must support real-time interoperability with operations, employees, customers, partners, and suppliers. Therefore, 5G capabilities and 5G-enabled business models need to provide the following key capabilities [35, 37]:

- Purpose-based reliable connectivity, coverage, and capacity.
- Efficiency for all mission- and business-critical communications.

- Security and safety.
- Low latency, traffic prioritization, and the ability to enable rapid endpoint communications.
- Agility to rapidly deploy and monetize new services and/or reduce operations costs.
- Interoperability and integration of IoT.

From a standardization point of view, in Release 17, private-network support is being further extended by introducing support for neutral host models, where the network owner and service provider need not be the same entity. This includes enablers for accessing standalone private networks using credentials from third-party service providers, including public network operators. Furthermore, support for onboarding and provisioning of user equipments (UEs) to access private networks is being introduced [43]. Releases 17 and 18 enhance the capabilities of new radio – unlicensed (NR-U) through the use of the 60 GHz millimeter wave bands, as well as through new URLLC, IAB, and Sidelink capabilities. It has been seen that NR-U is particularly beneficial for industries operating a private network when licensed or shared bands are not available or provide insufficient capacity. Based on these new enhancements and emerging business opportunities, the traditional business models are changing because many anticipate that some of the most interesting and important applications for 5G will be in vertical industrial sectors and other private network applications. The Section 1.4 describes some selected examples from the industry's vertical-specific requirements.

1.4 Key Drivers of 5G in the Industrial Verticals

As the Fourth Industrial Revolution is underway, enterprises across all industries are looking for ways to embrace data-driven operations, adopt zero-touch automation, and transform the way people and machines work together. Their goal is to digitalize their operations to make production processes safer, more productive, and more sustainable. Connectivity is the key to achieving this goal, but the wireless networks at most industrial sites were not designed to connect all industrial devices or transform data into practical insight. They also cannot interface with legacy environments and enable mission-critical processes that demand universal broadband coverage, strong security, 24/7 availability, and deterministic performance.

Digital technologies are transforming global industries and disrupting traditional business models. One key driver of digital transformation is 5G, which will create a basis for new digital services, business models, and ecosystems related to digital transformation across industries [44]. Although the role of 5G-enabled business models is crucial across industries, research into this essential topic is relatively limited. Therefore, this chapter presents the key requirements and the key drivers of 5G enabling enterprises to reach their business objectives on the multiple industrial verticals. The case studies across industrial domains show that 5G-driven private wireless networks create a new basis for new, data-driven business models that disrupt the current models. The results show that rapidly increasing numbers of wireless networks across industries will unlock significant potential for new business models and digital services and drive overall digital transformation across industries. However, next-generation wireless connectivity technologies are needed to enable further the shift to a digital economy and thus realize the productivity and social benefits that a successful transition promises [45]. Massive and exponential (IoT) data generation is pushing network capacity to its limits and at increasing cost. In addition, network sustainability and secured data flow are important factors. There is an urgent need to find solutions that dramatically slow down the exponential growth of capacity demand, energy consumption, and related costs for a sustainable network. A solution is needed that enables secure and seamless data flow from device to cloud and back. Although cloud providers have developed excellent security for their IoT offerings, many organizations have security concerns and prefer methods where their sensitive data stay inside the walls of their organization.

Advanced technologies such as 5G have great potential to increase production productivity significantly through digital transformation [46]. 5G capabilities can help increase output results per unit of time that can be realized, for example, using a higher level of equipment utilization and through more precise process control. In addition, the multiple sensor systems can be used to monitor and manage processes and, for example, create automated alerts ensuring the set target level. This also helps increase the speed of turnaround when mission-critical processes can be automatically controlled. Finally, productivity can be improved by adopting and implementing industrial agile principles to remove obstructions in the process, to increase flexibility in adapting to change, and to improve the process continuously. A key component of this potential lies in the collaboration between stakeholders in the manufacturing and mobile ecosystem industries who have acted in parallel in the past. In the future, the 5G technology will have a major impact on industry and mobility and will enable manufacturers to complete end-to-end automation with the virtual deployment of new product lines or an entire factory. For example, 5G enables low-latency transfer of sensor data to and from many robots, including in non-wired locations, enabling usage of sensor data to and from machines for cloud robotics [46]. In addition, the low latency of 5G enables remote control of assets such as mobile working machines and drones. Furthermore, enabling digital platforms should provide industry users with multi-technology connectivity, industrial edge computing, industrial applications, and IoT devices.

Today, manufacturing systems go beyond simple connections to communicate, analyze, and use collected information to drive further intelligent actions. It represents an integration of IoT, analytics, additive manufacturing, robotics, artificial intelligence, advanced materials, and augmented reality [7, 27, 47]. However, it has been suggested that, in order to implement 5G in a manufacturing environment, collaboration across all systems, spanning corporate IT and the manufacturing OT, should be taken into account to capture the full potential of 5G capabilities [32]. Traditionally, automation and control systems, such as supervisory control and data acquisition (SCADA), and distributed control systems (DCS), are often referred to as OT. These systems are used to manage critical infrastructures, for example, within the process industry. Previously, these OT systems have had a degree of physical separation from IT infrastructures. Today, due to industry-specific requirements and new data-driven opportunities, these two technology environments are starting to converge and are more interconnected [48], opening up new business opportunities for digital services and industrial applications.

To realize these opportunities during a time of increasingly uncertain economic outlook and to respond to new opportunities, many industries have initiated a process of digital transformation. Many organizations believe that they are well on their way to achieving their digitalization goals, but the reality can be different, especially across organizations with industrial sites and facilities. Achieving the last 25–50% of a digitalization process is often the hardest, frequently focusing on more complex OT use cases that are left until last – but it can also be the most rewarding in terms of long-term benefits [49]. This is particularly true in verticals such as manufacturing, logistics, and utilities where there is a growing requirement for locations such as factories, utility plants, and logistics hubs to be ever more efficient. In verticals such as manufacturing, logistics, and the energy sector, many enterprises are discouraged from embarking on digitalization projects at these sites because they believe that modernizing premises involving a complex mixture of next generation and legacy equipment is prohibitively difficult and expensive. However, a well-crafted Industry 4.0 transformation should combine next-generation technologies, such as private 4G/5G, industrial edge computing, devices, and analytics, with existing machinery and wireless solutions (e.g. IoT) and, in doing so, provide a path to digitizing legacy systems. This approach can accelerate the delivery of secure smart industry solutions in a way that offers a genuine return on investment, particularly when they are delivered as part of an end-to-end managed or comanaged solution.

References

1 M. Porter and J. Heppelmann, "How smart, connected products are transforming competition," *Harvard Business Review*, vol. November, pp. 65–88, 2014.

2 F. Calvino, C. Criscuolo, L. Marcolin and M. Squicciarini, "A Taxonomy of Digital Intensive Sectors," OECD Science, Technology and Industry Working Papers, 2018.

3 J. Collin and A. Saarelainen, Teollinen Internet, Helsinki: Talentum, Alma Media, 2016.

4 P. Drucker, Management: Tasks, Responsibilities, Practices, New York: Harper & Row, 1973.

5 J. Collin, K. Hiekkanen, J. J. Korhonen, M. Halen, T. Itälä, M. Helenius, "IT Leadership in Transition – The Impact of Digitalization on Finnish Organizations," Aalto University Publication Series SCIENCE + TECHNOLOGY 7/2015, Espoo, pp. 29–34, 2015.

6 OECD, "OECD compendium of productivity indicators 2019," 2019. https://www.oecd-ilibrary.org/industry-and-services/oecd-compendium-of-productivity-indicators-2019_b2774f97-en.

7 H. Kagermann, J. Helbig, A. Hellinger and W. Wahlster, "Recommendations for Implementing the Strategic Initiative INDUSTRIE 4.0: Securing the Future of German Manufacturing Industry. Final Report of the Industrie 4.0 Working Group," 2013.

8 J. Pellikka, J. Collin and J. Penttinen, "Key Drivers of Industrial 5G for Industry Digitalization," in *2023 9th International E-Conference on Advances on Engineering, Technology and Management*, Submitted.

9 W. de Paula Ferreira, F. Armellini and L. De Santa-Eulalia, "Simulation in industry 4.0: A state-of-the-art review," *Computers & Industrial Engineering*, vol. 149, p. 106868, 2020.

10 P. Laiho, 5G-enabled Digital Transformation in the Finnish Forest Industry, Espoo: Master's Thesis, Aalto University, 2023.

11 K. Nosalska and G. Mazurek, "Marketing principles for Industry 4.0 – A conceptual framework," *Engineering Management in Production and Services*, vol. 11, no. 3, pp. 9–20, 2019.

12 I. Vuksanović, V. Kuč, V. M. Mijušković and T. Herceg, "Challenges and driving forces for Industry 4.0 implementation," *Sustainability*, vol. 12, no. 10, p. 4208, 2020.

13 A. Calabrese, M. Dora, N. L. Ghiron and L. Tiburzi, "Industry's 4.0 transformation process: How to start, where to aim, what to be aware of," *Production Planning & Control*, vol. 33, no. 5, pp. 492–512, 2020.

14 H. Kagermann and W. Wahlster, "Ten years of Industrie 4.0," *Sci*, vol. 4, no. 3, p. 26, 2022.

15 M. Herman, T. Pentek and B. Otto, "Design Principles for Industrie 4.0 Scenarios: A Literature Review," Working Paper, 2015.

16 M. Hermann, T. Pentek and B. Otto, "Design Principles for Industrie 4.0 Scenarios," in *49th Hawaii International Conference on System Sciences (HICSS)*, Koloa, HI, USA: IEE, pp. 3928–3937, 2016.

17 Industry IoT Consortium, "Home – Industry IoT Consortium," www.iiconsortium.org (online).

18 OECD, "Going Digital in a Multilateral World," 2018https://www.oecd.org/site/videos/2018/documents/C-MIN-2018-6-EN.pdf (online).

19 E. Commission, "Digital Transformation," 2019.

20 G. Vial, "Understanding digital transformation: A review and a re-search agenda," *The Journal of Strategic Information Systems*, vol. 28, no. 2, pp. 118–144, 2019.

21 G. Westerman, D. Bonnet and A. McAfee, Leading Digital: Turning Technology into Business Transformation, Boston, MA: HBR Press, 2014.

22 K. Warner and M. Wäger, "Building dynamic capabilities for digital transformation: An ongoing process of strategic renewal," *Long Range Planning*, vol. 52, no. 3, pp. 326–349, 2019.

23 R. K. Saha, P. Saengudomlert, C. Aswakul, "Evolution toward 5G mobile networks – A survey on enabling technologies," *Engineering Journal*, vol. 20, no. 1, pp. 87–119, 2016.

24 S. K. Rao and R. Prasad, "Impact of 5G technologies on Industry 4.0," *Wireless Personal Communications*, vol. 100, pp. 145–159, 2018.

25 T. Kumar, M. Liyanage, I. Ahmad, A. Braeken and M. Ylianttila, "User Privacy, Identity and Trust in 5G. A Comprehensive Guide to 5G Security," pp. 267–279, 2018.

26 J. Hemilä and J. Salmelin, "Business Model Innovations for 5G Deployment in Smart Cities," in *Proceedings of the ISPIM Innovation Summit, Melbourne, 10–13 December 2017.*

27 M. Attaran, "The impact of 5G on the evolution of intelligent automation and industry digitization," *Journal of Ambient Intelligence and Humanized Computing*, vol. 14, pp. 1–17, 2021.

28 GSMA, The 5G Era: Age of Boundless Connectivity and Intelligent Automation, GSM Association, 2017.

29 Allied Market Research, Industrial 5G Market by Component, End User, Enterprise Size, and Communication Type: Global Opportunity Analysis and Industry Forecast, 2021–2030, Allied Market Research, 2021.

30 B. Insight, "Industrial automation and wireless IoT," M2M Research Series, 2019.

31 World Economic Forum, The Impact of 5G: Creating New Value Across Industries and Society, 2020.

32 A. Aijaz, "Private 5G: The future of industrial wireless," *IEEE Industrial Electronics Magazine*, vol. 14, no. 4, pp. 136–145, 2020.

33 A. Mahmood, S. Abedin, T. Sauter, M. Gidlund and K. Landernäs, "Factory 5G: A review of industry-centric features and deployment options," *IEEE Industrial Electronics Magazine*, vol. 16, no. 2, pp. 24–34, 2022.

34 S. Zhang, "An overview of network slicing for 5G," *IEEE Wireless Communications*, vol. 26, no. 3, pp. 111–117, 2019.

35 S. Yrjola, "Technology antecedents of the platform-based ecosystemic business models beyond 5G," in *2020 IEEE Wireless Communications and Networking Conference Workshops (WCNCW)*, pp. 1–8, Seoul, South Korea, 6–9 April 2020.

36 P. Ahokangas, M. Matinmikko-Blue, S. Yrjölä, V. Seppänen, H. Hämmäinen, R. Jurva and M. Latva-aho, "Business models for local 5G micro operators,"

IEEE Transactions on Cognitive Communications and Networking, vol. 5, no. 3, pp. 730–740, 2019.

37 I. Lee, "The Internet of Things for enterprises: An ecosystem, architecture, and IoT service business model," *Internet of Things*, vol. 7, p. 100078, 2019.

38 J. Schneir, A. Ajibulu, K. Konstantinou, J. Bradford, G. Zimmermann, H. Droste and R. Canto, "A business case for 5G mobile broadband in a dense urban area," *Telecommunications Policy*, vol. 43, no. 7, p. 101813, 2019.

39 M. Cave, "How disruptive is 5G?," *Telecommunications Policy*, vol. 42, no. 8, pp. 653–658, 2018.

40 M. Matinmikko, M. Latva-aho, P. Ahokangas and V. Seppänen, "On regulations for 5G: Micro licensing for locally operated networks," *Telecommunications Policy*, vol. 42, no. 8, pp. 622–635, 2018.

41 J. Pellikka and P. Malinen, "Business models in the commercialization processes of innovation among small high-technology firms," *International Journal of Innovation and Technology Management*, vol. 11, no. 2, p. 1450007, 2014.

42 M. Johnson, C. Christensen and H. Kagermann, "Reinventing your business model," *Harvard Business Review*, vol. 86, no. 12, pp. 50–59, 2008.

43 3GPP, *Release 17 Update from SA2,* 2021.

44 J. Pellikka and T. Ali-Vehmas, "Managing innovation ecosystems to create and capture value in ICT industries," *Technology Innovation Management Review*, vol. 6, no. 10, pp. 17–24, 2016.

45 M. Graham and W. H. Dutton, Society and the Internet: How Networks of Information and Communication are Changing Our Lives, Oxford University Press, 2019.

46 I. Rodriguez, R. Mogensen, A. Schjørring, M. Razzaghpour, R. Maldonado, G. Berardinelli, R. Adeogun, P. Christensen, P. Mogensen, O. Madsen and C. Møller, "5G swarm production: Advanced industrial manufacturing concepts enabled by wireless automation," *IEEE Communications Magazine*, vol. 59, no. 1, pp. 48–54, 2021.

47 I. Rodriguez, R. Mogensen, A. Schjørring, M. Razzaghpour, R. Maldonado, G. Berardinelli, R. Adeogun, P. H. Christensen, P. Mogensen, O. Madsen, and C. Møller, "5G swarm production: advanced industrial manufacturing concepts enabled by wireless automation," *IEEE Communications Magazine*, vol. 59, no. 1, pp. 48–54, 2021, doi: 10.1109/MCOM.001.2000560.

48 Gartner Research, "Market Guide for Edge Computing," 04 October 2022. https://www.gartner.com/en/documents/4019489.

49 Nokia Press Release, "Nokia and GlobalData Market Research Reveals Private Wireless Enterprise Drivers and Return on Investment Data," 08 December 2022. https://www.nokia.com/about-us/news/releases/2022/12/07/nokia-and-globaldata-market-research-reveals-private-wireless-enterprise-drivers-and-return-on-investment-data/

2

Green Digital Transition: New Standards for Sustainability

2.1 Introduction

Continuing concern about the world's natural resources has become a critical issue for all organizations, institutions, and governments. Additional attention has been paid to the opportunities and alternatives for how organizations can minimize their carbon footprint through twin transition across industries. The rollout of 5G and leading-edge fiber networks is the foundation upon which decarbonization can thrive. Through robust connectivity in the Industry 4.0 context, societies can accelerate digitalization via wider rollout of sensors, Augmented Reality (AR) and Virtual Reality (VR), cloud, and analytics to maximize the sustainable benefits of technology. Policies that encourage broadband adoption and the digital transformation of industry, that maximize available spectrum for connectivity, and that enable rapid deployment of digital infrastructure will help meet climate change goals.

Today, companies are increasingly challenged by their customers and other key stakeholders to improve their energy efficiency (EE), reduce business-related emissions, and, in general, accelerate efforts to make their business more sustainable. Therefore, it is essential to seek new ways to define, deploy, and improve a company's environmental and social performance while also satisfying the stakeholders' demands and improving the company's competitive advantage and public image [1]. To promote a sustainable future, for example, the European Union (EU) has defined objectives to promote research and development (R&D) and the use of digital technologies to pursue a green future through systematic transformations by introducing the concept of twin transition [2]. The term "twin transition" refers to an intertwined and simultaneous green and digital transition

5G Innovations for Industry Transformation: Data-Driven Use Cases, First Edition.
Jari Collin, Jarkko Pellikka, and Jyrki T.J. Penttinen.

Published 2024 by John Wiley & Sons, Inc.

to offset companies' carbon footprints, underlining the key role of 5G among other key digital technologies to enable companies and other organizations to reach their objectives on sustainability.

For example, the process industry has increased its focus on a wide variety of digital technologies to improve productivity, efficiency, and safety of their operations, while minimizing capital and operating costs, safety, and environmental risks. Based on the new digital capabilities and related technologies as part of Industry 4.0, we have already seen tangible developments in productivity, sustainability, and cost reduction across several industry sectors, including mining, pulp and paper, and oil and gas as well as in manufacturing, all driving digital transformation enabled by 5G and data-based digital solutions. Another example from the mining industry shows that twin transition can help companies to realize several targeted benefits in terms of enhanced safety as well as productivity by removing operators from hazardous environments and improving communication of mine operations, offering real-time insights. In addition, autonomous vehicle technology, Internet of Things (IoT) sensors and analytics, machine learning (ML), and artificial intelligence (AI) are the key building blocks for the next-generation mine. They have the potential to improve dramatically efficiencies, productivity, safety, sustainability, and asset predictability and, generally, optimize operational processes and costs. These technologies will accelerate multiple perspectives of mining operations from exploration to exploitation. Given the nature of most mining applications, one of the key supporting technologies for the digital transformation of mining is, specifically, 5G wireless connectivity. Current wireless technologies have not been engineered to provide sufficient coverage, reliability, security, service prioritization, and/or support for mobility and IoT. Business-critical mining applications require robust wireless connectivity with very high reliability, intrinsic security, and predictable performance. Operations cannot be prone to failure and stoppages. Automated haulage, for instance, requires wireless that supports mobility and ultrafast reactions, often across extensive terrain with a frequently changing footprint without breaks in coverage. Vehicle-mounted CCTV cameras and drones require true mobile broadband. Tele-remote operations require very low-latency connectivity. Critical person-to-person communications will need support for real-time two-way voice and video, and emerging IoT and analytics applications will need to support massive numbers of devices and sensors.

2.2 Industrial 5G-enabled Green Transition

Development of the complex 5G-enabled systems and applications needed for industrial usage requires the real-time analysis of data collected from various sources and processing in the distributed structure. Therefore, for example, Digital

Twin (DT) modeling has been shown to be a beneficial approach to drive productivity and sustainability through real-time production monitoring and optimization systems [3, 4]. In general, DTs can enable simulation and visualization of entities based on the advanced 3D model enabled by the collected data, e.g. from the industrial process. In addition, DTs collect data from industrial systems and then apply that data to digital models for real-time visualization and autonomous optimization and testing. DT is a combination of cloud computing, ML techniques, and increased computer power that has made the concept of integrating all data a viable reality [4]. DT is computer software that takes real-world data about a physical entity or system as inputs and produces outputs. Data are required for analytics, prediction, and automation in a DT. The management of DT models and data quality becomes essential for its successful operation. The main purpose of a DT is to gain insight into and predict the performance of a physical product, process, or piece of infrastructure.

Numerous advantages grow from the energy industry's adoption of DT technology, such as improved asset performance, higher profits and efficiencies, and less harmful effects on the environment [5]. It is also important to note that organizational investments in the environment can positively affect a company's competitive advantage within a given industry's marketplace [4]. Technological innovation assists firms to use their limited resources to foster their competitive advantage and improve their sustainable performance [5]. The empirical evidence suggests that Industry 4.0 tools can also positively influence a company's green process innovation and supply chain management. Due to the centrality of IoT to Industry 4.0 and the digitalization of companies, it is possible to assume that IoT should positively enhance green competitive advantage.

Industry 4.0 and digitalization have been identified as major contributors to driving productivity and sustainability across industries [6, 7]. In the Industry 4.0 context, the interconnected computers, sensors, and intelligent machines communicate with one another, interact with the environment, and eventually make decisions with minimal human involvement [7]. Industry 4.0 is commonly described as the use of emerging technologies (IoT, big data, and cloud computing, among others) to enable autonomous systems [8]. This is accomplished through self-organization and diagnosis, real-time monitoring, and optimization as well as the capacity of a system to learn and adapt according to environmental changes [9]. Therefore, it is expected that Industry 4.0 technologies can also drive sustainability across industries [1, 8]. At the general level, digitalization enables improvements in manufacturing and industrial processes in order to realize the benefits of productivity, resource efficiency, and waste reduction [10].

In this chapter, the term "sustainability" is used as defined by the United Nations, that is, as a movement for ensuring better and more sustainable well-being for all, including future generations, and aims to address the perpetual global issues of injustice, inequality, peace, climate change, pollution, and environmental degradation. The concept of sustainability in the Industrial 4.0 context challenges traditional approaches to problem-solving and demands more systemic ways of addressing change [1]. From this perspective, the concept of the triple bottom line (TBL) has been developed to provide a framework. This consists of three key elements:

1) Economic sustainability is related to long-term economic growth and profitability while preserving environmental and social resources [10]. Viewed from this perspective, the growth of economic capital should not be at the expense of the decrease in natural or social capital.
2) Social sustainability to foster development of human and societal capital to create communities where everyone is protected from discrimination and has access to universal human rights and basic amenities such as security or healthcare [11].
3) Environmental sustainability, which refers to the consumption of those resources that can be reproduced. In addition, environmental sustainability is mainly concerned with maintaining the earth's environmental system equilibrium, the balance of natural resource consumption and replenishment, and ecological integrity [10].

Current society expects a shift in value creation from pure economic benefits toward holistic sustainability, including the social and environmental perspectives. Therefore, economies must transform to meet economic, environmental, and social standards equally to ensure a comprehensive sustainable development, summarized in the approach of the TBL [12].

2.3 Benefits of Industrial 5G for Sustainability

Digitalization provides access to an integrated network of unexploited big data with potential benefits for society and the environment. The development of smart systems connected to the IoT can generate unique opportunities to address challenges associated with the United Nations Sustainable Development Goals (SDGs) strategically to ensure an equitable, environmentally sustainable, and healthy society. Digitalization can be defined as the process of converting physically collected information and knowledge into a computer-readable language. The benefits resulting from digitalization have contributed to the development of tools and sensors integrated into the IoT environment. The IoT is a robust network of physical objects connected over the internet through embedded sensors, software, and other technologies that enable interchange and collection of data [13]. The convergence of simultaneously developed technologies for real-time analysis, ML, and AI produces a massive amount of data. The high value of these generated massive datasets has not yet been fully exploited but has created unique opportunities to catalyze the transition to more efficient and sustainable industrial processes. Industrial 5G as part of digitalization in the industrial context can foster sustainability in several ways.

2.3.1 Improved Energy Efficiency

By enabling real-time monitoring and control of industrial processes, 5G can help reduce energy waste and improve EE [14, 15]. This can help lower greenhouse gas emissions and reduce the overall environmental impact of industrial operations. The potential of Industry 4.0 is thus remarkable for achieving sustainable industrial value creation across social, economic, and environmental dimensions by improving resource efficiency. In addition, it has been seen that 5G deployments in manufacturing, logistics, transportation, and consumer verticals could reduce carbon dioxide (CO_2) emissions by 20 gigatons by 2030 [16]. In manufacturing, a single smart factory using 5G for predictive, preventative, and remote maintenance, as well as the deployment of automated guided vehicles (AGVs), is expected to reduce CO_2 emissions by around 103 tons by 2030. Furthermore, an AGV running over a 5G network is expected to be 45% more productive than other AGVs, due to the robust handover of signals between different access points [16]. 5G can also support the integration of renewable energy sources into industrial operations, as it can enable real-time monitoring and control of renewable energy systems [17]. This can help reduce reliance on fossil fuels and lower greenhouse gas emissions.

2.3.2 Reduced Downtime

Predictive maintenance practices enabled by 5G can help reduce downtime, as issues can be detected and addressed before they cause significant problems. This can lead to improved equipment performance and reduce the waste of resources, including energy and raw materials. The objective of smart manufacturing is to find opportunities to automate operations and use data analytics to optimize manufacturing performance. Smart manufacturing is recognized as a particular application of industrial IoT (IIoT). Implementations include installing sensors in machines to gather information and data on their performance and operational status. Before, the data were kept in databases, which were local to individual devices and utilized only to evaluate the reason for equipment failures after they occurred. Currently, by analyzing the information streamed off a whole utility's machines, operators can monitor the signs for failure in specific parts. This allows preventive maintenance to avert unplanned downtime of equipment. For example, the water industry is currently facing challenges due to climate change, urbanization, and aging infrastructure that make water security, safety, and sustainability key public policy concerns. Another example of the benefits is related to 5G-enabled predictive maintenance by using IoT and analytics in the oil and gas industry. Leveraged pervasive wireless coverage to collect data from IoT sensors can be used to feed asset management and advanced data analytics applications that can help avoid situations proactively that may lead to unexpected shutdowns of the industrial processes. With the advent and adoption of low-powered IIoT sensors, increased information, and operational data promise to revolutionize the way water systems are managed. The result is a powerful, holistic water management approach that can empower flexible and resilient community water management systems for cities and for different industry sectors.

2.3.3 Enhanced Safety

5G can also improve safety in industrial settings, as it can support the use of real-time data and information to manage decision-making and improve safety processes [14]. Additionally, greater automation of industrial processes enabled by 5G can reduce the risk of human error and improve overall safety outcomes. Positioning technology can track mobile and portable devices connected to the 5G network, accurately determining their locations where no global navigation satellite service coverage is available, for instance, in factories, warehouses, or underground facilities. As part of the factory test, an enhanced private 5G network was able to determine the precise position of assets such as AGVs, mobile robots, and mobile control panels – tracking their movements throughout the plant in real time. Precision localization is important for many applications in industrial environments, such as robot navigation, asset tracking, and worker safety. In addition, realizing both high-performance connectivity and high-accuracy positioning within a single private network's infrastructure also has many operational benefits, such as reducing the complexity of IT infrastructure, leading to a lower total cost of ownership (TCO) and higher returns on investments. For example, in the mining industry, 5G enables a wide range of use cases to improve safety such as situational awareness that, together with real-time mixed reality capabilities, enables 360° visibility of people, assets, infrastructure, and health and environmental conditions where it can save lives, prevent productivity losses, and increase operational efficiency. Other concrete benefits in the mining industry are the modern AR/VR systems that can be used to train professionals and give them real-time information. With step-by-step instructions delivered using AR glasses, maintenance staff can fix problems faster. In addition, personal smart devices and wearables can be used with geo-fencing applications to alert workers as they approach no-go zones. These capabilities can also provide a much-needed lifeline in confined spaces and other situations where the risk of accidents is high.

2.3.4 Improved Logistics and Supply Chain Management

Over the past few years, the advances in the development of 5G, IoT, and cloud technologies for industrial use cases have promoted transition from traditional logistics and supply chain management toward Industry 4.0. These technologies can connect multiple assets, mobile machines, and other essential resources. The real-time tracking and monitoring of products and materials enabled by 5G can improve supply chain management and reduce waste [18]. By optimizing transportation and logistics, 5G can help reduce emissions from transportation and improve the efficiency of supply chains. 5G capabilities can be also used as an integrated entity in the value chain by collecting, sharing, and utilizing data to provide real-time information on the different steps within supply chain. This enables decision-makers to manage the process efficiently.

As described above, industrial 5G has the potential to play a key role in fostering sustainability by enabling industries to improve their operations, reduce waste, and lower their environmental impact. By leveraging the full potential of 5G technology, industries can help build a more sustainable future. According to [19], 82% of the companies surveyed stated that the most important factor, or one of the most important factors, in their strategy and decision-making processes is sustainability. There are many important reasons for this to be the case from both financial and commercial perspectives. Manufacturers, energy producers, natural resource extractors, and transport and logistics companies are particularly exposed on both sides, as organizations in these verticals are perceived as being a source of carbon emissions and/or at the forefront of the process toward reaching net zero. The key customers and other stakeholders are more likely than ever to make purchasing decisions based on their perception of a company where sustainability plays a significant role in how those perceptions are formed. In addition, the companies spoken to made it clear that rising energy costs means the need for efficiency is greater than ever. The enterprises surveyed revealed that making plants, logistics sites, and vehicle fleets more carbon neutral is their biggest barrier to reducing Scope 1 emissions (i.e. carbon emissions generated by internal operations). In particular, it is recording and processing sustainability data from sites and equipment that are hardest to achieve. This difficulty also impacts these organizations' ability to report their emissions data externally, which has regulatory implications and would have a negative impact when dealing with (potential) commercial partners.

5G may play a part in the transformation of the energy sector toward smart grids, smart metering, and digitalization of power plants. Other positive environmental impacts can be realized in the mobility sector by facilitating decarbonized multimodal and shared mobility as well as in production toward utilizing highly efficient smart factory and Industry 4.0 concepts. With such potential, highly environmentally conscious initiatives can accelerate and support 5G ecosystem performance. Industry 4.0 through digital transformation enabled by 5G can result in more resource-efficient industries that generate less waste and have greater productivity, a significant response to climate challenges, and also provide alternative consumption options for individuals and communities.

2.4 5G Radio Network and Energy Efficiency

In recent years, energy consumption has become a large area of focus for many fields, including communications engineering. This is mainly due to two reasons. First, companies and governments have set ambitious environmental goals with plans to become carbon neutral over the next few years due to the ongoing climate crisis. Second, the recent increase in energy costs has greatly impacted the operational expenditure of companies in the communications sector. According to Eurostat, the price of electricity for non-household consumers in the first half of 2022 increased by almost 50% on average in the EU when compared to the previous year [4]. These recent steep increases in electricity prices are due to exceptional circumstances, and prices are likely to decrease in the near future, but the events have shown the volatility of the energy market.

Around 80% of all electricity consumed in a mobile network is consumed by the radio access network (RAN), which is why EE efforts have traditionally been focused on that part of the network. This is no different for 5G, where RAN EE has been included in the 3rd generation partnership project (3GPP) specifications starting from the first 5G release of Release 15, although EE has been expanded to include the 5G core network as well in later releases. Starting from Release 16, [1] has defined the main key performance indicators (KPIs) for EE and intercell methods for energy savings in 5G. The two high-level KPIs both consider the energy consumption of a system but approach it from two different directions. The first KPI considers how much data can be moved in the network for a certain amount of energy consumed, and the second one considers what kind of coverage area can be achieved with a certain amount of energy consumption. The first KPI related to data volume is more useful in data-dense deployments found in many industrial environments, whereas the second gives an important metric for large coverage area deployments found in rural areas.

Two of the largest trends in 5G in terms of EE are the use of higher frequencies and massive multiple-input and multiple-output (MIMO) antennas. Massive MIMO not only plays a large role in the 5G promise of higher capacity, but it also includes the downside of higher energy consumption. This is because, with a higher number of MIMO layers, more power amplifiers (PAs) are required, which have been found to consume 50–80% of energy used by a RAN [2]. Some of this problem can be mitigated by powering multiple antenna elements using a single PA, but more PAs are still required when the number of MIMO layers increases massively. MIMO muting has been proposed as a possible solution to reduce the high energy consumption of massive MIMO for low data volumes. Some simulations have shown that powering down a part of the PAs for low data volumes could reduce energy consumption by as much as 30% [3].

A great deal of 3GPP EE standardization efforts are still ongoing. Release18, which is expected to be completed by 2024, will include studies on leveraging AL and ML and virtualization to increase EE. Especially interesting will be the new use cases that leverage the new network data analytics function (NWDAF) to provide data and analytics on which RAN and core network EE could be improved [4].

References

1 M. Ghobakhloo, "Industry 4.0, digitization, and opportunities for sustainability," *Journal of Cleaner Production*, vol. 252, p. 119869, 2020.

2 H. Birkel and J. Müller, "Potentials of Industry 4.0 for supply chain management within the triple bottom line of sustainability – A systematic literature review," *Journal of Cleaner Production*, vol. 289, p. 125612, 2021.

3 H. Nguyen, R. Trestian, D. To and M. Tatipamula, "Digital twin for 5G and beyond," *IEEE Communications Magazine*, vol. 59, no. 2, pp. 10–15, 2021.

4 T. Deng, K. Zhang and Z. Shen, "A systematic review of a digital twin city: A new pattern of urban governance toward smart cities," *Journal of Management Science and Engineering*, vol. 6, no. 2, pp. 125–134, 2021.

5 "ABI Insight," 2022. [Online].

6 S. Rehman, D. Giordino, Q. Zhang and G. Alam, "Twin transitions & Industry 4.0: Unpacking the relationship between digital and green factors to determine green competitive advantage," *Technology in Society*, vol. 73, p. 102227, 2023.

7 M. Shehab, I. Kassem, A. Kutty, M. Kucukvar, N. Onat and T. Khattab, "5G networks towards smart and sustainable cities: A review of recent developments, applications and future perspectives," *IEEE Access*, vol. 10, pp. 2987–3006, 2021.

8 R. Morrar, H. Arman and S. Mousa, "The fourth industrial revolution (Industry 4.0): A social innovation perspective," *Technology Innovation Management Review*, vol. 7, no. 11, pp. 12–20, 2017.

9 S. Yang, M. R. Aravind Raghavendra, J. Kaminski and H. Pepin, "Opportunities for Industry 4.0 to support remanufacturing," *Applied Sciences*, vol. 8, no. 7, p. 1177, 2018.

10 G. Tortorella and D. Fettermann, "Implementation of Industry 4.0 and lean production in Brazilian manufacturing companies," *International Journal of Production Research*, vol. 56, no. 8, pp. 2975–2987, 2018.

11 D. Corujo, J. Quevedo, R. Aguiar, P. Paixão, H. Martins and Á. Gomes, "An economic assessment of the contributions of 5G into the railways and energy sectors," *Wireless Personal Communications*, vol. 129, pp. 1–24, 2022.

12 J. Elkington, "Partnerships from cannibals with forks: The triple bottom line of 21st-century business," *Environmental Quality Management*, vol. 8, no. 1, pp. 37–51, 1998.

13 B. Wit and K. Pylak, "Internet of things (IoT) for implementation of triple bottom line to a business model canvas in reverse logistics," *Electronic Markets*, vol. 30, pp. 679–697, 2020.

14 G. Kavallieratos, S. Katsikas and V. Gkioulos, "SafeSec tropos: joint security and safety 899 requirements elicitation," *Computer Standards & Interfaces*, vol. 70, p. 103429, 2020.

15 N. Dempsey, G. Bramley, S. Power and C. Brown, "The social dimension of sustainable development: Defining urban social sustainability," *Sustainable Development*, vol. 19, no. 5, pp. 289–300, 2011.

16 J. Cheng, W. Chen, F. Tao and C. Lin, "Industrial IoT in 5G environment towards smart manufacturing," *Journal of Industrial Information Integration*, vol. 10, pp. 10–19, 2018.

17 A. Dolgui and D. Ivanov, "5G in digital supply chain and operations management: Fostering flexibility, end-to-end connectivity and real-time visibility through internet-of-everything," *International Journal of Production Research*, vol. 60, no. 2, pp. 442–451, 2022.

18 S. Rao and R. Prasad, "Impact of 5G technologies on Industry 4.0," *Wireless Personal Communications*, vol. 100, pp. 145–159, 2018.

19 Global Data, "Private wireless enterprise drivers and return on investment data," Global Data, 8 December 2022.

3

Smart, Connected Products with APIs Transform Industry Ecosystems

3.1 Introduction

Professor Michael Porter, who is generally regarded as the father of the modern strategy field, has influenced thinking on company strategy and competition since the 1980s. His influential books on five competitive forces [1], competitive strategy [2], competitive advantage [3], and many others have directed the way of company strategies across different industries over the last few decades. In recent years, his theories and thoughts on competition have been applied to smart, connected products of the digital era [4, 5].

The evolution of products into intelligent, connected devices – which are increasingly embedded in broader systems – is radically reshaping companies and competition [5]. Smart, connected products are physical products that contain smart components (e.g. sensors and actuators) and connectivity components (e.g. transmitters and receivers), and allow the use of digital cloud-based services in real time. Smart, connected products require companies to build and support an entirely new technology infrastructure [5]. This new "technology stack," which builds on the existing product management systems, consists of three layers: (1) product with embedded software and hardware to enable sensing, (2) connectivity to enable communication, and (3) product cloud to enable data analytics and smart applications. In addition, the stack includes a suite of security tools, a gateway for external information sources, and integration with enterprise business systems [5].

We believe smart, connected products not only transform competition and companies but also fundamentally change the way that new industry ecosystems are

5G Innovations for Industry Transformation: Data-Driven Use Cases, First Edition.
Jari Collin, Jarkko Pellikka, and Jyrki T.J. Penttinen.

Published 2024 by John Wiley & Sons, Inc.

created and developed. In this chapter, building on Porter's thinking, we investigate the ecosystem-wide opportunities and dynamics to utilize this kind of intelligent, and connected product system.

First, digital transformation involves changes across the whole industry ecosystem – not only inside a single company. Industry ecosystems need to seamlessly work together to provide end customers with real-time, data-driven services on top of traditional physical products. Second, these cyber–physical solutions require an integrated, secure connectivity and computing platform that is easily accessible for industry ecosystem partners globally. In our approach, industrial 5G represents the connectivity and computing platform. The third element consists of standard application programming interface (API) interfaces that are needed to enable the development of cloud-based applications and their usage anywhere in real time. Without these APIs, it simply becomes impossible to work effectively together in an ecosystem. They also guarantee the speed and scalability of the application development. We start with industry ecosystems.

3.2 Industry Ecosystems – Driving Digital Transformation

In a dynamic business environment, an organization's capability to catalyze the emergence and guide the development of an ecosystem can offer an increasing potential and powerful source of competitive advantage in the Industry 4.0 era [6]. When looking at ecosystems, it is essential to understand their differences compared to other collaboration models between different actors. Ecosystems differ from other models, for example, from the point of view of the following characteristics [6–9]. First, the ability of actors to produce value is dependent on other actors in the ecosystem, the measures they implement, and the changes caused by the measures in relation to other ecosystem actors. From this point of view, the ecosystem has often simultaneously represented both producers of products and services as well as their beneficiaries and users. Second, dependencies can appear between organizations operating in the ecosystem both horizontally and vertically, which separates them from, for example, traditional value and/or subcontracting chains. Third, ecosystems are dynamic in nature, in which case they must also be managed and developed, for example, the content and methods of their operations continuously.

Companies operating in ecosystems develop new assets and capabilities to drive the realization of the set business objective. At the same time, they strive to manage dynamic changes in the business environment, for example, by anticipating changes in the market and business environment at the ecosystem level. Cooperation models can be different in form and content, varying from, for example, product development partnerships to share-based joint ventures, cooperation in manufacturing, and joint sales and marketing arrangements.

The general goals of such cooperation also include combining mutual learning and talent development, access to new markets and technologies, and speeding up market introduction. Through the development of joint capabilities, companies can gain new and wider opportunities from digital transformation to create jointly, for example, new technological capabilities, innovation, solutions, and digital services to meet consumer needs and future business opportunities [7, 10, 11]. Ecosystems can aim, in particular, for cooperation arrangements (e.g. co-creation), through which companies combine their individual offerings into a broader, unified, and customer-oriented solution. With this solution, companies can cocreate value that they could not do alone. Studies have shown that ecosystems have become a new basis for collaboration to drive digital transformation across industries.

3.2.1 The Emergence of Industry Ecosystems

The concept of an ecosystem has its original roots in the field of biology, which was defined in the 1930s as a biological assemblage interacting with its associated physical environment and located in a specific place [12]. In 1993, James F. Moore introduced the concept of the business ecosystem and, by using biological metaphors, defined it as [13]: "Business ecosystem is an economic community supported by a foundation of interacting organizations and individuals – the organisms of the business world. The economic community produces goods and services of value to customers, who are themselves members of the ecosystem. The member organisms also include suppliers, lead producers, competitors, and other stakeholders. Over time, they coevolve their capabilities and roles, and tend to align themselves with the directions set by one or more central companies. Those companies holding leadership roles may change over time, but the function of ecosystem leader is valued by the community because it enables members to move toward shared visions to align their investments, and to find mutually supportive roles" [13].

The ecosystem as a business concept has gained ground as the business landscape has changed due to traditional industries collapsing and competition with rivals changing from producing products to creating better ecosystems. Knowing how to build healthy ecosystems and when to build them creates competitive advantage for a company [14]. Since the 1990s, business ecosystem thinking has become extremely popular in industries, governments, and research institutes leading to new types of "ecosystems," such as knowledge and innovation ecosystems [15]. The term "digital ecosystem" has also become ubiquitous around digitalization [16]. Originally, the concept of the industrial ecosystem was introduced in 1989 by [17]. As pointed out by [17], the industrial ecosystem would function as an analog of biological ecosystems when an optimal status may never be attained in practice. However, both manufacturers and consumers must change their habits to approach it more closely if the industrialized world is to maintain its standard of living. In this industrial ecosystem context, it is possible to develop a more closed ecosystem, which is more sustainable in the face of decreasing supplies of raw materials and increasing problems of waste and pollution. The ideal industrial ecosystem includes the optimized use of energy and material with minimized waste and emissions. However, in 1989, technology was not mature enough to enable this, something which has changed since then.

In this book, the term industry ecosystem refers to a dynamic business ecosystem within a specific industry that is in the middle of digital transformation impacting traditional ways of working, roles, and business models. Therefore, the ecosystem can be managed strategically if decision-makers understand the mechanisms underlying the ecosystem dynamics.

In general, ecosystems are organizational collectives combining capabilities and assets from the ecosystem members to create value offerings to a defined market and customer segment. Since the introduction of the concept of "ecosystem" in the business context [13], the concept has been studied from multiple perspectives covering, for example, strategic management [18], Internet of Things (IoT), and platforms [19–21]. In addition, it has been also indicated that ecosystems can be categorized as business ecosystems, innovation ecosystems, and knowledge ecosystems [22]. From this perspective, each ecosystem may have different objectives in order to create value for its members. For example, "business ecosystem" emphasizes the importance of the business benefits whereas "innovation ecosystem" focuses more on the innovation policies to drive regional development and entrepreneurial aspects [23]. In addition, "knowledge ecosystem" typically focuses on joint research and development (R and D) efforts to create new knowledge for members [22].

Compared to the other concepts of interorganizational collaboration, ecosystems can be distinguished from other community constructs through their participant heterogeneity, role of digitalization and platforms, type of system-level output, variety of participant interdependence, and nature of governance [6, 13, 24]. For example, digital service ecosystems, especially when organized around digital platforms, constitute an increasingly important driver of service innovation, as they facilitate a flexible combination of varied resources for mutual value creation through data and/or service exchanges. For example, data and their processing were viewed as an expense by organizations, necessary to business operations, but not as a potential driver for new business development and commercialization opportunities. Recently, new technologies such as artificial intelligence (AI) and big data analytics have shed light on the high potential value held by data, such as product-related data or customer data, and triggered the development of innovative smart services.

Ecosystems have fewer hierarchical arrangements and more independent participants, which pose challenges to business model creation and orchestration. For example, the decision-making principle means the mechanism and priority of the decision may be very different among actors in the ecosystem. Therefore, it is relevant to analyze whether ecosystem members behave in such a way that they contribute to an increased value of the focal participant's offering, and whether they can be persuaded to do so especially if there are fewer hierarchical arrangements. Working cooperatively with other players, such as private and public organizations, opens up new opportunities to use and build complementary assets to drive the organization's objectives further. This can be achieved, for example, by using novel ways created to facilitate resource mobilization and data sharing to develop new digital products or services through ecosystem orchestration [7]. Ecosystem orchestration is linked to the value offering that the ecosystem collectively produces, which inevitably arises from the participating organizations' business models. This has created a need for research that focuses on ecosystem business models, arguing that most business model conceptualizations to date overlook the systemic participation of diverse actors and overemphasize the role of a single organization [25]. An ecosystem business model presents a situation in which it is impossible for a single company to govern all relevant resources and activities needed for developing, producing, and marketing technology-based services [26]. This creates a need to consider a new business model logic that highlights the properties of ecosystem life cycles and strategies in the local context [27]. In order to realize these benefits, understanding the ecosystem life cycle and orchestration of the ecosystem are essential elements to ensure the realization of the value for the ecosystem members. In this chapter, we define the term "ecosystem orchestration" as "the set of deliberate and purposeful actions undertaken by the ecosystem orchestrator or a hub organization to plan, manage, and mobilize resources for co-creating value and to reach the set objectives" [8, 10, 28]. This means that an ecosystem must be able to define the key elements of the business model, the orchestration of these elements, and how to improve the content and the way of working continuously to create and capture value for members in a sustainable way.

However, the current understanding of ecosystem orchestration that addresses ecosystem emergence approaches and provides concrete perspectives for ecosystem orchestrators is relatively limited [28]. Current practices lack a theoretical foundation that addresses the development and change of innovation ecosystems over time and does not consider the inherent dynamics of 5G ecosystems that lead the phases within their life cycle: (1) conceptual design, (2) ecosystem building,

(3) operation and maintenance, and (4) succession [29]. Ecosystem business models are still incomplete, both in terms of how they create value for companies and how they are orchestrated or built for scalability (adapted from one industry context to another). Therefore, ecosystem business models that can drive digital transformation, productivity, sustainability, and new business development is a field requiring decision-makers' attention [9, 10, 26]. It has also been noted that it is challenging to orchestrate an ecosystem effectively that consists of multiple actors, assets, data, and resources [28] and, therefore, it is essential that an ecosystem must be able to identify in more detail the key value creation elements, drivers, roles, and key constraints [11].

3.2.2 Collaboration Benefits

There are clear benefits to driving productivity and sustainability objectives across industry verticals. Collaborating with other companies, academic institutions, and government organizations can provide access to new technologies and innovations that can help drive digital transformation. This can help industries stay ahead of the curve and adopt new technologies more quickly. Ecosystem partners can cocreate new digital services together with other vendors such as other service providers, manufacturers, integrators, communication service providers (CSPs), and other technology providers since all the required capabilities and domain-specific knowledge cannot typically be found in-house. On the other hand, process innovation is deeply entrenched in internal operations, and equipment suppliers depend on gaining access to their customers' knowledge so that they can customize process solutions to their idiosyncratic design requirements. Combining these assets the ecosystem partners can develop operational innovations faster and create extra value with new revenue streams. In addition, sharing of expertise and knowledge with the ecosystem partners improves understanding of the opportunities in a specific area, such as data analytics or cloud computing. This enables the value-adding elements for the customer to be defined, e.g. through co-creation when partners can collaborate closely to solve the defined challenge together [24]. This also helps to shorten time-to-market and, more generally, commercialization with a new solution. Obtaining some of the required capabilities (e.g. for research and development activities) from the business partners rather than building them in-house can help companies to reduce their financial commitments along with technology and market risks.

Co-innovating with the ecosystem partners can promote the emergence of new ideas, concepts, and assets based on the combination of both internal and external sources to create value for the targeted market and customer segments [30]. The core of co-innovation includes engagement, experience, and co-creation for value that is difficult to imitate by the competition outside of an ecosystem. Ecosystem collaboration can also provide opportunities for joint problem-solving and collaboration on projects. This can help industries tackle complex challenges and find solutions more quickly and effectively. By transferring and pooling their technological know-how and resources, companies can also enhance execution of co-creation activities [17, 31]. Therefore, ecosystem collaboration, especially in the industrial domain, can contribute to the customer experience. Collaborating with other organizations can also help improve the customer experience by providing access to new services and products that may not have been possible otherwise. In addition, ecosystems enable improvements in agility to develop new business opportunities and go-to-market channels as part of the commercialization process together with the ecosystem partners [32]. Finally, ecosystem collaboration can drive better efficiency and cost management. For example, sharing resources and infrastructure can reduce the costs related to R and D and commercialization of new innovation without high investments [31].

Overall, ecosystem collaboration can provide industries with the resources and expertise they need to succeed in their digital transformation efforts. By working together, organizations can achieve their goals more quickly and effectively, and create value for their customers and stakeholders. In summary, ecosystems and ecosystem collaboration can play a critical role in helping industries succeed in their digital transformation efforts by providing access to resources and expertise that may be difficult to obtain otherwise.

3.2.3 New Value Through Data Sharing

Currently, the business of numerous companies is solely based on data, be it data collection, transmission, processing, storage, etc. However, the strong and disruptive 5G dynamics could go much further than silo data processes (i.e. operated by a single party) and may lead to the emergence of new usages and new players, with the sharing of data among authorized stakeholders and the creation of combined datasets for even higher value-added services [33]. However, the sharing and combining of datasets holds a number of technical, business, and regulatory challenges, as well as new threats, such as private data leakage or cyberattacks. The question of trust should have a significant impact on the evolution of 5G ecosystems. For example, today, some very large industrial companies make the choice to invest massively in standalone nonpublic networks that are designed, deployed, and operated internally. Indeed, this deployment scenario is perceived as less risky when it comes to network isolation and security, even if this may not be always true from a technical perspective. Building trust between stakeholders is, thus, absolutely necessary to allow multi-tenant 5G solutions, as well as sharing of costs and liabilities. From the industrial 5G point of view, industry could capture value through data exchange.

Today, the data generated by the production lines, machines, and sensors on a shop floor are generally isolated, confined to a given business process, and available to the machinery owner only. However, data could be shared in an automated and secured manner between other industrial processes and among authorized stakeholders, to provide valued information on similar machinery or on a complementary domain for improved production, maintenance, and monitoring. For example, this would allow a manufacturer to fine-tune its production lines based on the data received directly from providers and related to the workpieces used on these production lines. Another example of such collaboration relates to logistics between suppliers and consumers, and intralogistics within a factory, on assembly lines. Indeed, delivering the right parts to the right place at the right time would significantly improve efficiency and productivity, while reducing stocks.

The platform provides an architecture of participation for the ecosystem constituents, thereby helping reduce costs and, through this, the digital platform reinforces shared institutional logics and streamlines rules of engagement and shared expectations in service exchanges, giving rise to a greater number and variety of these [23]. In addition, research has identified ecosystem offerings as malleable and users as having a broader range of opportunities to define the value offering compared to the context of conventional supply chains [28].

From the industrial 5G and industrial IoT (IIoT) perspectives, the benefits of ecosystem collaboration and ecosystemic business models may be significant. In addition, it has been seen that, in the IoT environment, understanding the business models of company partners is important for long-term success to drive twin transitions across industries. In the IIoT context, companies can enhance their knowledge base and data-driven business growth if they can effectively use their agility and adapt themselves to changes in the market and the current business environment. In addition, companies have the potential to gain from ecosystem collaboration and partnerships due to their ability to use shared knowledge, data, and assets efficiently [28]. In the IoT domain, business model innovations tend to cross multiple industries and drive digital transformation through ecosystems in which sensors, data, and smart objects facilitate business models and new digital services and applications. Such services and applications are more often based on collaboration between multiple ecosystem partners that are incrementally or radically novel in terms of their modularity or architecture [21]. At the same time, it has been seen that traditional firm-centric business models are not suitable to respond to the emerging needs driven by digital transformation and data-driven digital services. This also challenges the company-centric business models as product manufacturers may need to change their business models from the operation-centric models to more digital service-centric models.

3.3 Industrial 5G – Building a Platform for Industry Ecosystems

The penetration of digital platforms in the consumer markets is a good example of the rapid development of industry-wide ecosystems. The common platforms challenge the traditional value creation by driving horizontal and vertical interconnection in ecosystems. They enable ecosystem members to create and capture new value. In the industrial context, we define digital platforms as products, services, or technologies that provide the foundation upon which external organizations can develop their own complementary products, technologies, or services [7].

3.3.1 Digital Platforms and Platform Economy

Platforms are not a novel phenomenon: newspapers and shopping malls can be seen as platforms that bring together consumers and providers [34]. In addition, stock markets can be seen as early platforms [35]. However, technological revolutions have modified technological settings, which in turn have allowed new ways of value creation and capture to arise. What is different now with digital platforms is the fact that no more physical assets or infrastructures are needed as IT enables creation of platforms [34].

Nowadays, platforms are a complex combination of software, hardware, operations, and networks [36]. They provide a framework for building online applications, websites, and other digital solutions. These platforms typically offer a range of tools and services to developers and users, including data storage, security features, and integration with third-party services. Especially in the consumer markets, digital platforms have already been used for a wide variety of applications, from social media networks to e-commerce websites to mobile apps. Amazon Web Services (AWS), Google Cloud Platform, and Microsoft Azure are the most well-known examples of such digital platforms. One of the key benefits of digital platforms is their ability to facilitate collaboration and communication among users on a global scale. Many platforms include features that allow users to share information and data, communicate with each other in real time, and collaborate on projects. Digital platforms have become an essential part of running many business operations in several industries.

Platform economy is a business based on enabling value-creating interactions between external producers and consumers [37]. It is centered on digital platforms connecting buyers and sellers, service providers and customers, or any other combination of participants in an online marketplace. Platform economy is a set of online digital arrangements whose algorithms serve to organize and structure economic and social activity [36]. Digital platforms serve as intermediaries that facilitate exchanges between different parties, creating new forms of economic value. The platform provides an open, participative infrastructure for these interactions and sets governance conditions for them. The platform's overarching purpose is to complete matches between users and facilitate the exchange of goods, services, or social currency, thereby enabling value creation for all participants [37].

Platform economy has transformed the way we think about future business models and digital opportunities in industry ecosystems. The platform economy combined with ecosystems creates an emblematic organizational form of the digital age [38]. Competition in platform economy is more dynamic and complicated, but Porter's competitive forces still apply with some extra details [39].

The platform is at the core, emphasizing that platforms are a central agent at the nexus of a network of value creators [38]. This means that the platform provides infrastructure and rules to the marketplace that bring together consumers and providers.

The platform brings together the producers and consumers while keeping the providers and owners at the core of the platform. The role of the owners of the platform is to control IPs and governance, while providers are responsible for the interface that the end-user utilizes. On the other hand, producers create offerings and consumers use the offerings during the interactions on the platform. It is essential to note that participants who act as consumers or producers at one point can switch their roles at some other point while creating value for the platform.

The platform needs to attract both producers and consumers that create and consume value [40]. Three essential aspects of platform economy are: (1) the participants, producers, and consumers; (2) value unit, meaning the offering that is exchanged; as well as (3) some filters that help to match the right producers to the right consumers. There are typically two types of platforms that create value: transaction and innovation. Transaction platforms focus on creating matches for the participants on the platform while working as an intermediary [38]. Innovation platforms, on the other hand, allow third-parties to create additional services and products while the platform's role is to foster innovation [38].

It is crucial to understand how the transition to platform economy has changed the business landscape. Platform economy drives companies that have been production and service-oriented to change their business models as business models are no longer founded on ownership but rather on interactions with vast numbers of participants. This can be seen as innovation, and owning assets are not only activities carried out internally in companies [41]. At the same time, companies strive to cross traditional industry boundaries while co-creating value across industries [42], pp. 4.

A shift from a linear value chain to a horizontal platform structure has been influential in many industries. A linear value chain encompasses a structure where producers and customers are at separate ends of the line, and the value is generated step-by-step starting from the producer designing the product or service, manufacturing it, delivering it, and ending up with the customer buying the delivered product [37]. Platform structure consists of complex relationships of producers, consumers, and the platform itself, that consume, exchange, or cocreate value for all participants [37]. Thus, the source of value is the ecosystem that the platform creates [36].

The rise of platform economy is due to rapid technological development over the last few decades. Novel digital technologies allow the connection of individuals with other individuals and organizations with little friction, which is the premise of platform economy [38]. In addition, open interfaces give access to a company's internal information and resources for partners and customers which drives innovation and, thus, generates value for the platform [43].

3.3.2 Industrial 5G as a Digital Platform

From an industrial 5G perspective, many industrial companies are seeing new opportunities to drive digital transformation through digital platforms, since 5G capabilities, edge and cloud-based digital platforms, and related services support, for example, critical and massive machine-to-machine communication in order to improve productivity and quality management. In addition, 5G networks have been also defined as "connectivity-focused platforms" where a CSP plays the focal role as the platform owner [21, 28, 33]. From this perspective, industrial 5G capabilities and 5G networks can act as intermediaries, interconnecting various internal and/or external actors for purposes such as information and data sharing, product development, and supply and demand matching. Therefore, a 5G ecosystem can be decomposed into two main aspects [33]: (1) the network service provisioning aspect and (2) the vertical sector service consumption aspect (see Figure 3.1).

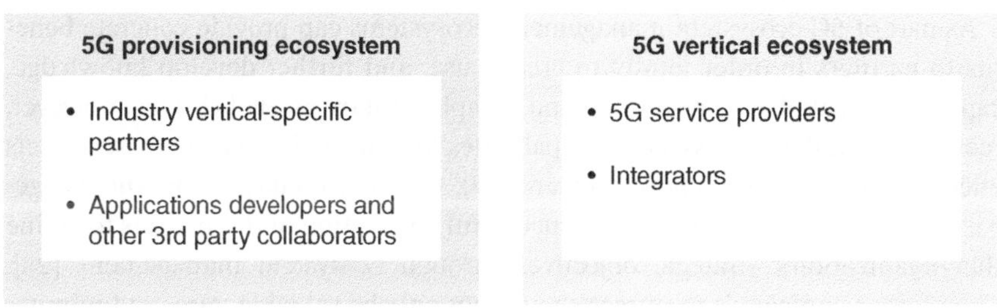

Figure 3.1 Categories of 5G ecosystems. *Source:* [33] / 5G Infrastructure Association.

The 5G provisioning ecosystem encompasses those roles and actors who take part in developing, delivering, and providing 5G services [33]. Traditionally, the telecom industry has been seen as a value chain where network operators source the resources necessary to provide fixed and mobile telecommunication services. The notion of a 5G provisioning ecosystem acknowledges an increased dependency on other roles and actors to grow the 5G market. The 5G vertical ecosystem blackboxes the 5G provisioning ecosystem and focuses on other actors who work closely together as part of vertical industries [9, 21]. While roles and actors from the telecommunication sector are still present in this ecosystem, the emphasis is on other roles that apply 5G services in their value creation and can be domain-specific.

The separation of aspects simplifies the discussion at each level by hiding the complexities inherent in each of the aspects. The involved actors, once they have concluded that it is attractive to engage in the ecosystem, must iteratively refine their strategies of positioning their firms for overall value creation, considering the growth of the ecosystem. Hence, the evolved strategies of the involved actors should leave room for growing the ecosystem by making it attractive for additional actors to become involved.

In the 5G provisioning ecosystem, the required services are mapped to roles that are expected to deliver these services [33]. This allows a mapping of the roles of the actors seeking to create value. The model includes all necessary providers, operators, and suppliers needed to deliver 5G services to the customers. The 5G provisioning ecosystem can be seen as a multi-actor platform ecosystem, as opposed to the single-actor platform ecosystems by the hyperscalers [8]. The challenges in evolving from 4G into 5G and beyond, and further developing such a multi-stakeholder ecosystem platform, require both technical and business coordination, development, and interoperability between the involved stakeholders. Similarly, the 5G vertical ecosystems are composed of other actors that assume roles necessary to adopt 5G services provided by the 5G provisioning ecosystem. In the context of 5G vertical enterprise customers, a large number of actors can assume complementary roles and are often competency-specific to the vertical sector they act in [13, 28].

As part of 5G ecosystem management, ecosystems can provide concrete benefits to partners in order jointly to create, use, and further develop knowledge, capabilities, data-based resources, and complementary assets [4]. This, however, requires the following assets: (1) capabilities to manage knowledge-based assets effectively (both internally and externally), (2) high quality of its knowledge-based assets (e.g. data), and (3) the successful application of these assets to define the organization's strategic objectives through ecosystem management [28]. Therefore, a company's resources should not only be valuable, rare, and inimitable to facilitate superior performance but the company must also have an appropriate strategy, organization, and processes in place to take advantage of the knowledge-based resources and data within the ecosystem [9, 11, 19]. In addition, it has been previously noted that the increasing role of data and data-based assets has become one of the primary drivers behind the digital transformation across industries enabling new innovation and digital services [9, 29].

This means that, for example, access and utilization of data-based assets among the ecosystem partners through connectivity and digital infrastructure is a critical success factor. A large majority of industry sectors and new business opportunities (e.g. IoT and digital healthcare) are generated by an ecology of private, public, and non-profit organizations, all often involved in innovation ecosystems. In order to stay relevant in the business, all ecosystem stakeholders (e.g. other startups, large companies, universities, and end-users) must strategize and continuously align both the inbound and outbound data and knowledge flows by using predetermined practices, e.g. through effective boundary resources including APIs and software development kits (SDKs) [8, 11]. Accessible interfaces (APIs) can be one value-creation element to accelerate value creation in an ecosystem. Accessible and open means that interfaces are not only publicly available but also the process of accessing them is made easy for all members of the ecosystem. For instance, firms who cocreate in the ecosystem should be invited into a smooth developer journey, and have access to SW developer kits, documentation, and developer communities.

In addition, reliable and solid technological enablement contributes to trust within the ecosystem and related communities and engagement among stakeholders in an ecosystem. The APIs, technological standards, or open-source SW are important self-reinforcing factors in an ecosystem [8, 19, 32]. First, their ease-of-use and wide distribution not only facilitate innovation by enabling recombination but also serve as a push for using them. Second, when large user groups become familiar with a specific interface and can learn from each other, the threshold for choosing a different interface is raised.

Furthermore, previous studies have shown that platforms and other enabling technologies and capabilities such as 5G, edge computing, and ML/AI can accelerate the value recognition function of absorptive capacity and, therefore, accelerate further diffusion of knowledge, data, and knowledge acquisition and co-development among the ecosystem members [11, 33]. These activities can also enable startups and small and medium-sized enterprises to deepen their specialization while further developing their business opportunities through business concepts, business models, market launching, and business planning [9, 44, 45]. The final vital element in the ecosystem is the need for at least one industry leader company or a "keystone" company. Their role is to ensure the continuous improvement of the ecosystem, encourage new innovative startups to join the ecosystem, and create offerings that are compatible with the expectations of other ecosystem stakeholders, including end-users. This role may coincide with the roles and activities taken on by the orchestrator(s) within the innovation ecosystem.

3.4 Standard APIs – Enabling Common Digital Platforms

An API is an interface that exchanges data and services between two computers [46]. It is a set of protocols, routines, and tools for building software applications. APIs define how different software components should interact, and allow for communication between different applications, services, and systems. More precisely, APIs can be defined as "the calls, subroutines, or software interrupts that comprise a documented interface so that a (usually) higher-level program such as an application program can make use of the (usually) lower-level services and functions of another application, operating system, network operating system, driver, or other lower-level software program" [47]. In addition to data exchange, APIs offer the possibility to allow different user applications to reuse services and functionalities [46]. This enables developers to create applications and services that can connect to other systems and exchange data.

APIs can be used for a variety of purposes, such as integrating with third-party services, sharing data between different applications, and automating routine tasks. It can be seen as a contract that enables communication and exchange of data between two or more applications to be executed over a network while using language both sides understand. The contract ensures that the API provider provides a specification of the API that the developer needs to agree on and follow to be able to use the API. The specification describes the functionality offered, availability, technical constraints as well as legal and business constraints related to the API. This makes the connection between the API provider and developer efficient as essential aspects are documented and consistent. The contract also increases confidence which in turn increases the use of API, as the developers acknowledge that they can rely on the API [48].

From a more technical perspective, an API is a group of protocols and definitions that allow two or more applications to communicate and exchange data with each other [49]. For the developer to be able to use the API, they need to make API calls which are processes of communicating with the API to retrieve from, or send data to, the API endpoint [49]. There are two main API protocols to make the calls: remote procedure call (RPC) and Representational State Transfer (RESTful) APIs (i.e. REST). Over recent years, GraphQL APIs have gained popularity due to their data querying capabilities [50] with public interest toward event-based APIs significantly increasing [46].

3.4.1 Types of API

There are different ways to categorize API types. A technical aspect is one way to classify them. RESTful APIs are the most common type. They use the HTTP protocol to transfer data and are based on a set of architectural principles for building web services. Another type is the Simple Object Access Protocol (SOAP) API that uses XML as the data format and typically requires more processing power and resources than RESTful APIs. GraphQL is a query language and runtime for APIs that provides a more efficient, powerful, and flexible alternative to traditional RESTful APIs. It enables clients to request only the data they need, rather than downloading all available data. Webhooks are not strictly APIs, but they do provide a way for applications to communicate with each other. Instead of requesting data, webhooks allow applications to receive notifications when certain events occur, such as when a new user signs up on a website. RPC APIs are a type of API that enable one program to request a procedure or function from another program, running on a different machine or in a different process. Each type of API has its own advantages and disadvantages, depending on the use case. Developers need to choose the right type of API based on the needs of the application, the resources available, and the desired performance.

Another way to classify APIs is based on their business purpose [35]. APIs are designed with a certain business arrangement in mind rather than concentrating on certain contents or applications [48]. This influences the architectural viewpoint of different types of API interfaces. In this book, we examine three types of APIs: public, partner, and internal [34, 49, 51]. Figure 3.2 defines the main differences between the API types.

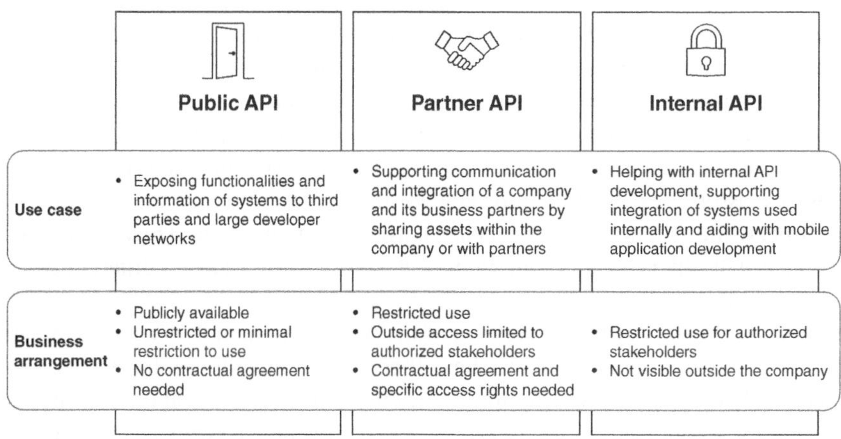

Figure 3.2 Main API types [35].

The use cases of public APIs are available to anybody, exposing functionalities and information of a company's system to third-parties. This means that there is usually no contractual agreement other than the terms of use for using the API. A partner API is often used to support a company's relationship with its business partners by enabling communication and integration [51]. This means that the business information is shared with its partners. The nature of the business arrangement is for restricted use, and outside access is only possible for authorized stakeholders [49]. Thus, partner APIs require a contractual agreement. Last, internal APIs are used inside companies mainly to support integration of different internal IT systems. Internal APIs are exposed only to a company's internal systems and applications while they are hidden from external users [49]. The nature of the business arrangement is for restricted use only by authorized stakeholders within the company.

APIs are becoming increasingly important in modern industry ecosystems. They allow different software applications and systems to communicate and interact with each other, creating a seamless integration of business processes and increasing efficiency. In addition to improving operational efficiencies and reducing costs, APIs also create opportunities for innovation and disruption in the industry ecosystem. Startups can leverage APIs to build new products and services, while established companies can create APIs to expand their reach and create new revenue streams.

3.4.2 API Value Chain

An API value chain defines the participants in the digital platform economy and helps the understanding of the motivations of different participants at different phases in the chain [35]. An API-enabled economy requires companies to open up their business models while cocreating value with their partners and developers. Therefore, understanding the motivations of different parties is essential for allowing businesses to execute API strategy properly [48].

An API value chain consists of five elements. The elements are business assets, API, developers, applications, and end-users. These elements describe how the business assets are converted into value for the end-user through an indirect API channel while generating value for each participant in the chain. The value proposition of API differs for different kinds of businesses as well as for different types of APIs [48].

Business assets are the first element of the API value chain. Business assets can be any assets that the business wishes others to use, including information, services, and products. A key factor for the API to be successful is that the asset needs to have some value to the end-user. The owner of the business assets needs also to understand what the value for them is at publishing the asset. Through internal and partner APIs, the owner of the business assets might want to publish business assets that they might not want to, or do not have the right to, publish outside the organization or partnership. The asset might still have some value as it could be used, for example, for making operational data more easily available for those who need it. On the other hand, through public APIs, the owner of the business assets might want the asset to reach a larger audience [35].

The second element of the API value chain is the API itself. The type of API can be either public, partner, or internal, but the key is that the intended audience should be able to use it easily [35]. The API provider is responsible for making that happen while exposing the business assets through the API. Especially in the case of public APIs, it is essential that the API provider creates an environment that fosters the use of the API in addition to promoting it for developers. Most of the time, the API provider is the same entity as the owner of business assets regardless of the API type. However, if the API provider is not the same entity as the owner of the business assets, establishing agreement for redistribution for reward is needed.

Developers form the third element of the API value chain. After the API has been published, it is hoped that some developers use it and create some applications or other products with it [35]. These developers can be, for example, product managers who are leading teams of developers internally at the company, business analysts, or individual developers. In the case of internal APIs, the developers are usually some employees of the company. With partner APIs, the developers can be internal developers or some partners. With public APIs, the developers can basically be anyone. The motivation of these individual developers differs vastly. They might be interested in experimenting with new technology, fostering innovation, carrying out public services or activism, or earning money through making applications.

Applications are the fourth element of the API value chain. After the developers have used the API, they might create some applications that use some of the business assets provided by the API. As the application is finished, it needs to be published to the market for it to bring value. Examples of distribution channels to market include App Stores as well as marketing assistance. However, it is important to acknowledge the differences in distribution of applications that are developed based on internal, partner, and public APIs. Applications created based on internal and partner APIs can be used either internally in the company, within partnerships, publicly, or all of these. On the other hand, applications created based on public APIs are usually available to everyone. Either way, it is essential to promote the application toward the intended end-users.

The last element of the API value chain is end-users. They are the entities that will use the created application and, thus, the business assets provided through the API. An end-user can be an entity of a certain company, belong to a partnership, or be an individual end-user. Through the usage of the application, the end-user should receive some benefits and value. At the same time, this provides value for the developer and API provider as well as the owner of business assets.

All in all, understanding the different motivations of all participants in the API value chain is essential to be able to execute a good API strategy [48].

References

1 M. E. Porter, How competitive forces shape strategy, UK: Macmillan Education, pp. 133–143, 1979.

2 M. Porter, Competitive Strategy, New York: Free Press, 1980.

3 M. Porter, Competitive Advantage, New York: Free Press, 1985.

4 M. Porter and J. Heppelmann, "How smart, connected products are transforming competition," *Harvard Business Review*, vol. November, pp. 65–88, 2014.

5 M. Porter and J. Heppelmann, "How smart, connected products are transforming companies," *Harvard Business Review*, vol. October, pp. 97–114, 2015.

6 A. Gawer and M. Cusumano, "Industry platforms and ecosystem innovation," *Journal of Product Innovation Management*, vol. 31, no. 3, pp. 417–433, 2014.

7 M. Brettel, N. Friederichsen, M. Keller and M. Rosenberg, "How virtualization, decentralization and network building change the manufacturing landscape: An Industry 4.0 perspective," *International Journal of Information and Communication Engineering*, vol. 8, no. 1, pp. 37–44, 2014.

8 R. C. Basole and J. Karla, "On the evolution of mobile platform ecosystem structure and strategy," *Business & Information Systems Engineering*, vol. 3, no. 5, p. 313, 2011.

9 J. Pellikka and T. Ali-Vehmas, "Managing innovation ecosystems to create and capture value in ICT industries," *Technology Innovation Management Review*, vol. 6, no. 10, pp. 17–24, 2016.

10 N. Foss, J. Schmidt and D. Teece, "Ecosystem leadership as a dynamic capability," *Long Range Planning*, vol. 56, no. 1, p. 102270, 2022.

11 A. Attour and N. Lazaric, "From knowledge to business ecosystems: emergence of an entrepreneurial activity during knowledge replication," *Small Business Economics*, vol. 54, no. 2, pp. 575–587, 2020.

12 A. Tansley, "British ecology during the past quartercentury: The plant community and the ecosystem," *Journal of Ecology*, vol. 27, pp. 513–530, 1939.

13 J. F. Moore, "Predators and prey: A new ecology of competition," *Harvard Business Review*, vol. 71, no. 3, pp. 75–83, 1993.

14 J. Moore, The Death of Competition: Leadership and Strategy in the Age of Business Ecosystems, HarperBusiness, 1996.

15 K. Valkokari, “Business, innovation, and knowledge ecosystems. How they differ and how to survive and thrive within them,” *Technology Innovation Management Review*, vol. 5, no. 8, pp. 17–24, 2015.

16 M. Krivý, Digital Ecosystem: The Journey of a Metaphor, Digital Geography and Society, 2023.

17 K. Möller, S. Nenonen and K. Storbacka, “Networks, ecosystems, fields, market systems? Making sense of the business environment,” *Industrial Marketing Management*, vol. 90, pp. 380–399, 2020.

18 O. Dedehayir, S. Mäkinen and J. Ortt, “Roles during innovation ecosystem genesis: A literature review,” *Technological Forecasting and Social Change*, vol. 138, pp. 18–29, 2018.

19 A. Hein, J. Weking, M. Schreieck, M. Wiesche, M. Böhm and H. Krcmar, “Value co-creation practices in business-to-business platform ecosystems,” *Electronic Markets*, vol. 29, no. 3, pp. 503–518, 2019.

20 Y. Cai, B. Ramis Ferrer and J. Luis Martinez Lastra, “Building university-industry co-innovation networks in transnational innovation ecosystems: Towards a transdisciplinary approach of integrating social sciences and artificial intelligence,” *Sustainability*, vol. 11, no. 17, p. 4633, 2019.

21 S. Leminen, M. Rajahonka and R. W. M. Wendelin, “Industrial internet of things business models in the machine-to-machine context,” *Industrial Marketing Management*, vol. 84, pp. 298–311, 2020.

22 E. Autio, S. Nambisan, L. Thomas and M. Wright, “Digital affordances, spatial affordances, and the genesis of entrepreneurial ecosystems,” *Strategic Entrepreneurship Journal*, vol. 12, no. 1, pp. 72–95, 2018.

23 E. Autio and L. Thomas, “Tilting the playing field: Towards an endogenous strategic action theory of ecosystem creation,” *World Scientific Reference on Innovation Volume 3: Open Innovation, Ecosystems and Entrepreneurship: Issues and Perspectives*, vol. 3, pp. 111–140, 2018.

24 K. Ranjan and S. Read, “An ecosystem perspective synthesis of co-creation research,” *Industrial Marketing Management*, vol. 99, pp. 79–96, 2021.

25 H. Wieland, N. N. Hartmann and S. L. Vargo, “Business models as service strategy,” *Journal of the Academy of Marketing Science*, vol. 45, pp. 925–943, 2017.

26 T. Palo and J. Tähtinen, “A network perspective on business models for emerging technology-based services,” *Journal of Business & Industrial Marketing*, vol. 26, pp. 377–388, 2011.

27 E. Stam and A. Van de Ven, “Entrepreneurial ecosystem elements,” *Small Business Economics*, vol. 56, no. 2, pp. 809–832, 2021.

28 J. Pellikka and T. Ali-Vehmas, “Fostering techno-entrepreneurship and open innovation practices in innovation ecosystems-the case of Nokia,” in *Handbook of Research on Techno-Entrepreneurship*, Edward Elgar Publishing, pp. 175–197, 2019.

29 M. Virtanen and J. Pellikka, "Integrating the opportunity development and commercialisation process," *International Journal of Business and Globalisation*, vol. 20, no. 4, pp. 479–496, 2018.

30 E. Von Hippel, S. Ozawa and J. De Jong, "The age of the consumer-innovator," *MIT Sloan Management Review*, vol. Fall, pp. 27–33, 2011.

31 M. Jacobides, C. Cennamo and A. Gawer, "Towards a theory of ecosystems," *Strategic Management Journal*, vol. 39, no. 8, pp. 2255–2276, 2018.

32 P. Graça and L. Camarinha-Matos, "Performance indicators for collaborative business ecosystems – Literature review and trends," *Technological Forecasting and Social Change*, vol. 116, pp. 237–255, 2017.

33 5G-IA, "5G Infrastructure Association Vision and Societal Challenges – Working Group Business Validation, Models, and Ecosystems Sub-Group," 2021.

34 M. Arajärvi, M. Saukonoja and P. Vänttinen, "Julkisen hallinnon API-periaatteet," 2022. [Online].

35 E. Koivula, "Building Competitive Advantage with API Strategy – Case Study of Established Enterprises," Master's Thesis, Aalto University, 2023.

36 M. Kenney and J. Zysman, "The rise of the platform economy," *Issues in Science and Technology*, vol. 32, no. 3, pp. 61–69, 2016.

37 G. Parker, V. Alstyne, C. Marshall and P. Sangeet, Platform Revolution: How Networked Markets are Transforming the Economy and How to Make them Work for You, New York: W. W. Norton & Company, Inc., 2016.

38 A. Gawer, "Digital platforms and ecosystems: Remarks on the dominant organizational forms of the digital age," *Innovation: Organization and Management*, vol. 24, no. 1, pp. 110–124, 2022.

39 M. Van Alstyne, G. Parker and S. Choudary, "Pipelines, platforms, and the new rules of strategy," *Harward Business Review*, vol. April, 2016.

40 J. Moilanen, M. Niinioja, M. Seppänen and M. Honkanen, API-talous 101, Alma, 2018.

41 A. Gawer. Digital platforms and ecosystems: remarks on the dominant organizational forms of the digital age. *Innovation*, vol. 24, no. 1, pp. 110–124, 2022.

42 T. Ojanperä and T. O. Vuori. Platform strategy: transform your business with AI, platforms and human intelligence. Kogan Page Publishers, 2021.

43 T. Ojanperä and T. Vuori, Platform Strategy: Transform your Business with AI, Platforms and Human Intelligence, New York: Kogan Page Inc., 2021.

44 I. Oskam, B. Bossink and A. de Man, "Valuing value in innovation ecosystems: How cross-sector actors overcome tensions in collaborative sustainable business model development," *Business & Society*, vol. 60, no. 5, pp. 1059–1091, 2021.

45 R. Adner, "Ecosystem as structure: An actionable construct for strategy," *Journal of Management*, vol. 43, no. 1, pp. 39–58, 2017.

46 L. E. A. Vaccari, Application Programming Interfaces in Governments: Why, What and How, EU, 2020.

47 M. Shnier, Dictionary of PC Hardware and Data Communications Terms, Sebastopol, CA: O'Reilly Media, Inc., 1996.

48 D. Jacobson, G. Brail and D. Woods, APIs: A Strategy Guide, O'Reilly Media, Inc., 2011.

49 J. Davidse, "API Fundamentals," 2020. [Online]. Available: https://developer.ibm.com/articles/api-fundamentals/?mhsrc=ibmsearch_a&mhq=api. [Accessed 25 May 2023].

50 T. Mikkonen, R. Klamma and J. Hernández (eds.), "Web Engineering: 18th International Conference, ICWE 2018. Lecture Notes in Computer Science, vol. 10845, Springer, 2018. doi: 10.1007/978-3-319-91662-0.

51 M. Boyd, "Private, Partner or Public: Which API Strategy Is Best For Business?," 2014. [Online]. Available: https://www.programmableweb.com/news/private-partner-or-public-which-api-strategy-best-business/2014/02/21. [Accessed 10 April 2023].

4

New Capabilities of 5G SA

4.1 General

4.1.1 5G Evolution Toward Standalone

The third-generation partnership project (3GPP) Release 15 defines technical specifications (TS) of the 5G system (5GS), including 5G new radio (NR) and 5G Core (5GC) architectures. It defines a full set of the new 5G features and capabilities, such as network slicing and software-defined networking (SDN). It also defines intermediate architecture options that pave the way for the full 5GS, combining the previous 4G radio access and core technologies with the 5G NR and 5GC additions. Figure 4.1[1] summarizes these options.

3GPP Release 15 defines the initial 5GS architecture and functionalities and describes the evolved Mobile Broadband (eMBB) data service, subscriber authentication and authorization, application support, edge computing, IP multimedia subsystem (IMS), and interworking with 4G and possibly other access systems. Some of the key differentiators of the 5G networks are network virtualization (decoupling the functions from the underlying hardware), network slicing (enabling creation of multiple networks on a common physical infrastructure), open source (opening doors for new ecosystem stakeholders), and edge (data centers).

5G gNB (gNodeB, next-generation Node B) radio access nodes form a 5G radio access network (RAN), interconnecting a set of 5G user equipment (UE) and 5GC architecture. UEs can have many forms such as smart devices and sensors.

1 Please note that option 6 was one of the candidate deployment options initially evaluated by the 3GPP, but it was removed from the final list. However, the numbering of the options remained, which explains the gap between options 5 and 7.

5G Innovations for Industry Transformation: Data-Driven Use Cases, First Edition.
Jari Collin, Jarkko Pellikka, and Jyrki T.J. Penttinen.

Published 2024 by John Wiley & Sons, Inc.

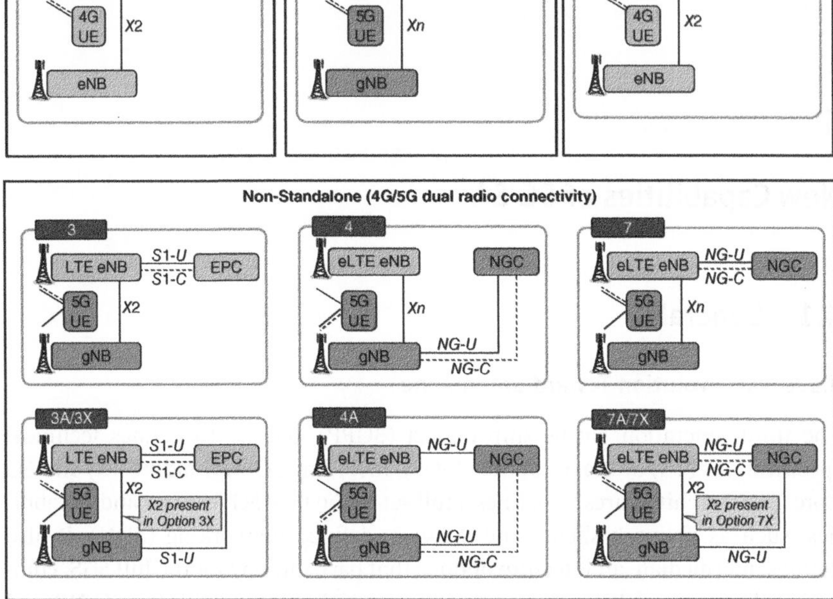

Figure 4.1 The 5G deployment options as defined by the 3GPP.

The consumer devices are typically voice-centric, supporting both voice service and data transmission, whereas data-centric devices include, for example, data hot spots, USB dongles, and IoT devices. As defined in 3GPP TS 38.300, a gNB node can provide 5G NR user plane (UP) and control plane (CP) protocol terminations toward the UE, and connect to the 5GC via the next generation (NG) interface.

A gNB is a logical component consisting of the radio equipment, such as baseband unit, transmitter, and receiver, connected to their respective antenna systems. The gNB forms radio cells with selected radio frequency (RF) bands. The cells can be physically close to the equipment, with coaxial cable connections to the antennas (the traditional manner), or fiber optic connections to a remote radio head (RRH) unit that is close to the antennas (currently the most common way to eliminate coaxial cable). In practice, a gNB can be generalized as a 5G base station although its logical protocol components can be distributed over larger physical areas, according to the 3GPP 5G RAN split model for the antenna and radio units (AU and RU), as well as distributed and centralized units (DU and CU, respectively).

Thus, a gNB refers to the radio functions that form a set of radio cells, whereas the physical base station is the equipment shelter housing the transmitters, receivers, power supplies, and other devices required to provide radio connectivity for the mobile devices. There are diverse kinds of shelter, ranging (for example) from a simple box installed on a wall or pole, through a room in a building, to a separate construction. As the 5G specifications allow the splitting of gNB functions, some may reside remotely in virtual containers of a data center. This is one of the key benefits of 5G.

4.1.2 Non-Standalone Options

Before full 5G deployment, operators may want to expedite deployment of its light version by relying on existing 4G infrastructure. This reutilization eases the deployment of 5G elements, which can be combined with the existing infrastructure as such, or by upgrading their supported features. A benefit of this is that the already deployed 4G system and its RF coverage can remain while the new 5G users can already start experiencing the benefits of 5G, such as increases in capacity and data speeds, in the same areas. These intermediate deployment options (3, 4, and 7) are non-standalone architectures (NSA) with combinations of 4G and 5G RAN elements.

In practice, option 3 is the most popular initial 5G deployment strategy as it allows operators to start providing "light-weight" 5G services while the technology and markets mature enough to deploy the more complete set of services and exploit the improvement of performance enabled by SA technology. In this option, the 5G UE relies on both 4G and 5G RAN components connected to a sole 4G evolved packet core (EPC). The benefit of this option is that its existing EPC does not even require awareness of the new 5G radio signaling interface, and the previously deployed long-term evolution (LTE) base stations (eNB, evolved NodeB) can be upgraded as anchors to interconnect the 5G gNB elements and 4G EPC. In this manner, the user data can be delivered between the EPC and eNB, or alternatively, also the gNB to increase the data speed.

There are several variants of option 3. In option 3A, both gNB and eNB deliver data to the EPC, but only the eNB has both user and signaling planes connected to the EPC. The data connectivity between gNB and EPC is provided by an already existing S1-U (4G) interface. In option 3A, there is no X2 interface between eNB and gNB, so these RAN elements aggregate data in an isolated manner.

Alternative option 3X includes an X2 interface for the eNB and gNB interconnectivity. This option can converge the traffic flow at the 5G gNB, routing part of it toward the 4G eNB while the 5G NR takes care of most of the traffic. This divided approach is useful if the 5G NR RF coverage is low or insufficient, in which case the split mechanism can offload more traffic to the eNB so that it is delivered via a better-performing RF channel. Thus, the 3X option can optimize the bandwidth using the X2 interface.

Option 4 can be used in the further evolution of the network toward full 5G. In this option, the 5G NR has more responsibilities than in option 3. The 5G gNB elements are anchors for connecting both the data and signaling directly to the EPC, and they aggregate the data flows from the further evolved 4G eNB elements (eLTE eNB) to the new 5GC.

Option 7 (including variants 7A and 7X) is like option 3, in that the 4G eNB elements handle both user data and signaling. The difference is that the core network (CN) is now based on 5G. In option 7, the 4G eNB elements still work as primary aggregators, but they are upgraded to deliver optimal 5G performance, and similarly to option 4 they are solely eLTE eNB elements.

In addition to these architectural deployment models, the operators may consider combining the models and using different options in different regions. An important optimization task is to decide the division of the frequency bands between the 4G and 5G RANs. One option is to collocate the same frequencies to NSA 5G RAN and 4G RAN elements that are installed at the same physical site to create similar low- and mid-band radio coverage areas (below 1 and 1–7 GHz) to ensure seamless user experience upon handovers. This strategy can also facilitate gradual transition to the new technology, enabling operators to reduce 4G eNBs while increasing 5G gNBs so that the 4G eNBs can fill coverage gaps, e.g. for less demanding IoT communications.

Another possibility is to collocate the eNB and gNB elements partially on different frequencies. In this way, 4G eNB cells can provide basic low- and mid-band coverage, whereas NSA 5G can serve customers on higher mm-wave bands to provide high-speed data and low latency. The 5G equipment can be collocated at existing 4G base station sites, but deployment of additional gNBs will be needed to fill the inevitable gaps due to the higher attenuation characteristics of the higher frequency bands.

The third scenario is 5G Standalone-focused as per the option 2. The 5G gNBs can be operated on any available licensed frequency band, with low-bands providing extensive coverage for low-speed traffic, mm-wave bands providing the highest data rates but very limited radio coverage, and mid-bands providing sweet spots to balance the coverage and data speeds. So, ideally, the deployment will combine all these band categories.

More information on the deployment options and scenarios can be found in Annex J of the 3GPP technical report (TR) 23.799 and 3GPP TR 38.801. The latter summarizes aspects of noncentralized, 4G/5G cosited, and centralized deployment, as well as shared RAN environments and heterogeneous deployment.

4.1.3 SA Deployment Options

Option 1 refers to the previous 4G systems, so it only includes 4G RAN components: the E-UTRAN radio access elements (eNB, evolved Node B) that connect the UP and CP of 4G UEs and 4G CN elements. E-UTRAN is a standard term, but in practical parlance, the 4G RAN is often referred to as LTE (more fully LTE mobile technology). EPC is the standard term for the 4G CN, sometimes called in practice system architecture evolution (SAE network, technology, or architecture). E-UTRAN and EPC form the complete 4G system, called evolved packet system (EPS).

Option 2 represents the full standalone 5G model, and is typically the ultimate current goal of 5G operators seeking the highest possible performance and most complete set of new features. It consists of solely 5G radio and 5G CN components and their respective sets of new features.

Option 5 is another standalone mode, but differs from option 2 in that it has a solely 4G core (EPC) connected solely to 5G NR gNB elements. It differs from the other intermediate options, which are all NSA. This is a deployment option that could be appropriate, for instance, in scenarios where home operators prefer to maintain a 4G user base, but at the same time provide inbound roamers with opportunities to use 5G-based features and services such as network slicing.

4.2 Technical Foundations of the SA Mode

4.2.1 New Capabilities

The 3GPP-defined 5GS incorporates various new concepts, including (among others) split CP and UP, network slicing, and service-based architecture (SBA). All these are essential building blocks for compliance with the strict ITU-R international mobile telecommunications 2020 (IMT-2020) requirements, and network slicing is a key enabler supporting multiple different use cases and instantiations of the same functionalities.

The 5GS provides enhancement of 4G performance, with evolution of its technical foundations and new features. Along with the intermediate NSA deployment options that allow operators to reutilize their 4G core, the clearest benefit for the consumer comes from eMBB, as the 5G gNB components can use wider bands, including the new mm-wave bands, which provide higher data speed connectivity. Thus, together with other 5G features, such as evolved MIMO antenna systems, the NSA options provide platforms for enhanced user experiences.

Deployment models involving the 5G CN can offer new functions, but to fully comply with the ITU-T IMT-2020's functional and performance requirements for 5G, a combination of new 5G gNBs and 5GC is needed.

4.2.2 Service-Based Architecture

The 5GC takes advantage of SBA [1]. The functional architecture of 5G allows fluent evolution of the implemented technologies and the functions can be replaced at optimal times. 5G also paves the way for more open multi-vendor environments, where UP and CP functionalities can evolve in parallel but independent fashions, enabling flexible deployment and variable network configurations via deployment of network slices (NS). The 5G CN houses all the functionalities required for establishing, maintaining, and releasing data and voice calls, and establishing any other required communication links such as messaging between the users and network's own signaling.

The new 5G incorporates significant upgrades to the previous CN architectures. Before 5GC, specific network functions required dedicated elements, such as subscription authentication or user registration, so each element was an individual component with its own hardware and software. Instead, 5G is based on virtualized network functions (VNF). Each function is provided by a set of software instances that can be run on virtualized hardware that can share its computing resources with other entities within containers. This means that 5GC can be deployed and operated in the cloud environment of the operator or separate third party data center.

3GPP Release 15-based 5G features suffice to deploy the new generation network, according to 3GPP TS 23.501, TS 23.502, and TS 23.503 specifications for the overall 5G architecture models and principles in roaming and non-roaming scenarios. Release 16 introduces further evolved features and enhancements so that the 3GPP specifications comply with the ITU IMT-2020 requirements for 5GSs. 5G can thus provide augmented performance and capacity compared to any of the previous generations through network virtualization. 5G also facilitates the emergence of innovative new business models and opportunities in associated ecosystems by defining open interfaces and open software concepts.

To comply with the strict requirements for ultra-low latency and high bandwidth, network functions (NFs) and content must be closer to the subscriber. This is facilitated by multi-access edge computing (MEC) that allows accelerated delivery of content, services, and applications increasing their responsiveness from the edge. The user experience will also be enriched by the highly capable network and service operations, with deeper insights into the radio and network conditions. In the 5G native environment, the applications are software instances that run on a virtual infrastructure that may be common hardware, including commercial off-the-shelf equipment provided its processor performance is adequate.

Network functions virtualization (NFV) and software defined networking (SDN) help reduce infrastructure and management expenses significantly so the communication between network entities moves from proprietary protocols to standard IP-based mechanisms, such as IPsec.

The 3GPP 5G SBA is applicable between the CP NFs of the CN. The 5G NFs can store their contexts in data storage functions (DSF) in a flexible way, which in turn increases the system's performance via optimized resource utilization. The new architectural model also supports enhanced access and mobility management function (AMF) resilience and load balancing, by allowing a set of AMFs within the same NS to handle procedures of any UE. The service-based 5G architecture refers to the network's capability to present its elements as NF in virtualized environments, which enhances its performance and provides means for new features.

Furthermore, 5G includes an extended quality of service (QoS) functionality, which can better differentiate data flows based on varying priority levels of services. It is based on existing QoS functionalities of the LTE system but adds new definitions.

Yet another benefit of 5G is that it provides support for various access systems. In addition to the new 5G radio network itself (next-generation RAN, NG-RAN), the 5G CN can also serve generic access networks (ANs) such as Wi-Fi (WLAN) hotspots. According to Release 15 definitions, the 5G CN can be interconnected with both 3GPP NG-RAN and 3GPP-defined untrusted WLAN networks and Release 16 includes more access options.

The SBA connects the required NFs via service-based interfaces (SBI). 3GPP Release 15 defines the first set of mandatory and optional NFs and their interfaces, while Release 16 and beyond specify additional NFs and SBIs that provide further evolution and enhancement of the 5GS's capabilities and performance. 5G NFs perform tasks as described in the 3GPP TS 23.501.

The SBA provides gradual 5G network deployments, and mobile network operators (MNOs) can exploit the latest advances of the virtualization concept as the specifications develop and evolved equipment becomes available.

The SBA enables presentation of the network elements as NFs that provide services to other (authorized) NFs via SBIs. The network repository functions (NRF) allow NFs to discover the services offered by other NFs.

Figure 4.2 depicts an example of the 5G SBA in a roaming scenario. The SBA consists of the NFs that work as instances on common hardware. The NFs include unified data management (UDM: a data repository for subscriptions comparable to the home location register and home subscription server in previous generations). Others include the policy control function (PCF), network exposure function (NEF), and non-3GPP interworking function (N3IWF) for interconnectivity to Wi-Fi hotspots. Customers connect to the network (visited or home) via the AMF and session management function (SMF) that are required for access and signaling. The routing of the data between gNB and external data networks is handled by the user plane function (UPF). For the roaming

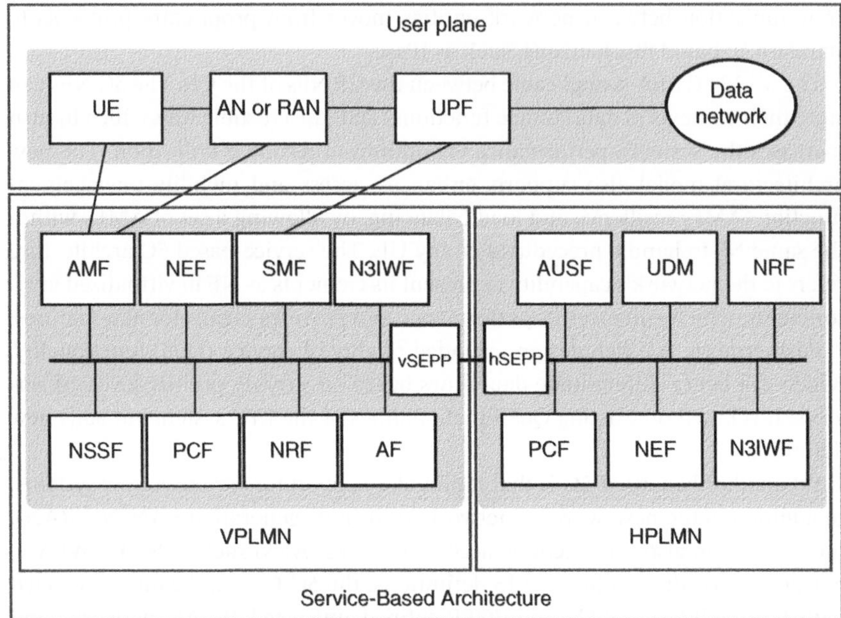

Figure 4.2 Example of the 5G SBA in a roaming scenario.

scenarios, the security edge protection proxy (SEPP) is the key component interconnecting the signaling path between home and visited networks either directly or via a roaming hub provider. For data connectivity, the UPF with assistance (as of Release 16) of the IPUPS (Inter-PLMN UP security) function interconnects the roaming operators.

In practical network deployment, the UPF represents a major cost compared to the set of SBA-NFs. This is an important consideration for any organization considering installing a private network as it provides various options for network sharing, and division of the NFs and UPF, and the costs of ownership vary accordingly.

4.2.3 Control and User Plane Separation

The 3GPP has defined the TS to allow a fluent transition phase from 4G to 5G services involving a change in the signaling of the original 4G EPC, via CUPS of the EPC. The transition splits the 4G SGW (service gateway) into SGW-CP and SGW-UP (SW-CP and UP, respectively) and the packet data network gateway (PGW) into GW-CP and PGW-UP [2].

CUPS (separation of functionalities of the SGW, PGW, and TDF) is defined in 3GPP Release 14 as paving the way for gradual 5G adaptation by reutilizing the already deployed 4G EPC infrastructure. This allows operators to consider flexible network deployment and operation via distributed or centralized deployment and independent scaling between CP and UP functions [2].

CUPS reduces latency in application services relying on UP nodes closer to the RAN so there is no impact on the number of CP nodes. The CUPS concept also supports increases in data traffic as the service utilization increases as the UP nodes can be added into the mobile network operator (MNO) infrastructure without impacting the numbers of SGW-C, PGW-C, and TDF-C elements of the mobile network. It also allows operators to add and scale the EPC node CP and UP resources independently, and thus independent evolution of the CP and UP functions. CUPS also enables SDN for optimized UP data delivery.

The 3GPP has further developed the user and CP split model in 5G. 3GPP TS 38.401 defines the 5G NR architecture and respective interfaces including the logical separation of signaling and data transport networks. In addition to the 5G infrastructure itself, this separation of the signaling and user data connections can be extended to the entire system, so that the user data could go through the 5G radio network and the signaling load could be diverted via the 4G radio, forming a certain type of extended carrier aggregation feature.

The 5G RAN and 5GC functions are separated from the transport functions. For example, the addressing scheme of the 5G RANs and 5G CNs are not tied to the addressing schemes of the transport functions, so the NG-RAN completely controls the mobility for an RRC connection.

The 5G UP transfers data via the packet data unit (PDU) session resource service from one service access point (SAP) to another [3]. The respective RAN and CN protocols provide the PDU session resource service. The 5G radio protocols for both UP and CP are defined in the 3GPP TS 38.2*xx* and 38.3*xx* series, and the 5GC protocols in the TS 38.41*x* series.

4.2.4 Network Slicing

One of the key functions of the 5GC is the Network Slicing enabled by the SBA. In this book, NS has been used as an abbreviation for Network Slice whereas Network Slicing is a functional term and does not have a dedicated abbreviation.

The 5G UE connects to the 5G infrastructure via the *Uu* radio interface. While traditional cellular network architectures only define uniform physical cellular networks, 5G network slicing provides means for a single MNO to form a set of parallel public land mobile networks (PLMNs) within the same physical area. In other words, slices are comparable to "networks within a network" that can be tailored in terms of performance to satisfy the varying needs of verticals.

A NSP, either an MNO or third party, can set up and optimize each slice individually to optimize the performance for specific usage scenarios. The operator can create and terminate slices dynamically as needed for required periods. For example, an operator can set up a specific slice to provide fast data for its users while another slice may serve a huge amount of low bit-rate sensors within the same service area. With adequate adjustment of the functional and performance parameters, different users can benefit from selection of the most suitable slices. This enhances the user experience and increases the MNO's possibility to optimize the offered network capacity.

The benefits of NS include the possibility for operators to provide dynamic, customizable end-to-end operations that meet specific needs of highly versatile users, including consumers and new verticals, such as enterprises engaged in the drone, virtual reality/augmented reality, and autonomous vehicle sectors.

For example, further development of artificial intelligence (AI) and machine learning (ML) technologies (among others) will ease service orchestration throughout the slices' life cycles, from concept testing, through commercial operations and maintenance to the end of the product.

The GSMA has produced a generic network slice template (GST) that presents attributes for key services relying on NS, referred to as service profiles in 3GPP TS28.541. The NS type, NEST, is defined in GSMA PRD NG.116.

In TS 28.530 the 3GPP defines a logical network that provides set of network capabilities and network characteristics to a specific NS, composed of RAN, CN, and transport network (TN) segments. It is composed of RAN, CN, and TN segments. NS enables the operation of one or more logical, customized networks on a shared common infrastructure in such a way that each slice meets service level agreements (SLA) with connected verticals.

The end-to-end NS spanning the whole chain of communication, covering UE, RAN or other AN, CN, TN, and network management system (NMS) supplied by various equipment vendors.

For diverse industrial verticals and other organizations, network slicing provides their customers, clients, or associates opportunities to subscribe to a slice using common infrastructure so that it delivers desired connectivity and other requirements.

Along with this possibility, the 5G SA system can adapt to the needs in a highly dynamic manner and is designed to cope with use cases, instead of offering rather uniform performance for each customer like previous generations.

We would prefer:

Network slicing involves individuals and organizations with the following roles:

- Communication service customers (CSCs), users of communication services, such as end users, tenants, and verticals;
- Communication service providers (CSPs) who offer, design, deploy, and operate communication services that may (or may not) be NS-based;

- Network operator (NOP) is responsible for the design, deployment, and operation of network services and for offering them to users;
- Network slice customers (NSCs), including CSPs or CSCs who use NSs as a service;
- Network slice providers (NSPs), including CSPs or NOPs that provide NSs as a service.

Through the diversity of this set of roles, NS extends the applicable business models and stakeholders beyond those associated with the legacy systems. Thus, the MNO can cooperate with third parties to manage the NSs.

For a NSP to set up a set of NSs it is important to understand the needs of their expected users and corresponding technical requirements that map to attributes (existing and extended GSTs) and values (NESTs). As outlined in GSMA PRD NG.130, interpreting users' needs can be challenging due to differences in terminology of the mobile communications ecosystems, standard-setting organizations, and verticals. To set up the GSTs and NESTs it is thus important to ensure that the NS stakeholders can interpret the practical requirements of the field correctly, understanding key current and future needs that are expressed in various non-standardized terms.

4.2.5 3GPP Network Exposure

The 5G networks are evolving to extend the business opportunities of the operators to include technical capabilities to offer services beyond network connectivity. A key element of such provision is the NEF.

The 3GPP SA6 working group considers mission-critical applications, and since Release 15 the group has developed solutions for new vertical applications within the 3GPP ecosystem. Their activities also include cross-industry promotion of the adoption of 3GPP 5G technology and features, such as a common application programmable interfaces (API) framework (CAPIF), service enabler architecture layer (SEAL) for verticals, and application layer support for selected services.

4.2.5.1 CAPIF

The CAPIF enables a unified northbound API framework across 3GPP network functions and is applicable to 4G and 5G networks. The aim of this initiative is to ensure a single approach, as described in 3GPP TS 23.222 [4], TS 33.122 [5], and TS 29.222 [6]. The CAPIF northbound APIs have been developed by 3GPP SA (service and system aspects) groups SA2 (for system architecture and service), SA4 (for multimedia codecs, systems, and services), and SA6 (for application enablement and critical communication applications). The aims are to ensure compliance of future 3GPP northbound API development with the CAPIF and provision of a single entry point for API invokers, i.e. vertical applications, toward the CAPIF APIs, including onboarding, discovery, authentication, and authorization.

The later releases define a further evolution of CAPIF further (eCAPIF) enabling third party API providers to leverage the CAPIF framework, the support for interconnection between two CAPIF providers, and the federation of CAPIF functions to support distributed deployments.

The CAPIF has been designed to cope with multiple APIs relevant to the 3GPP system's northbound communication to avoid duplication and inconsistency between different API specifications. The development of a common API framework facilitates the use of common services such as authorization as it includes common aspects applicable to any northbound service APIs.

TS 23.222 presents the APIs and functions needed to support the CAPIF, including the core, exposing, publishing, management, and invoker functions. Briefly:

- A third party application provider normally provides the API invoker. The third party has a service agreement with the PLMN operator. The same trust domain as the PLMN operator network may house the API invoker.
- The CAPIF core function supports various capabilities, including authentication of the API invoker and service API discovery, service API access policy provisioning, API routing information, API topology hiding information, and publishing and storage of the service API's information.
- An API exposing function (AEF) is the provider of the service APIs and service communication entry point of the service.
- An API publishing function (APF) enables the API provider to publish the service APIs' information to enable the discovery of service APIs by the API invoker.
- An API management function (AMF) enables the API provider to administrate the service APIs.

Figure 4.3 presents the basic functional model of CAPIF, which can be extended to cover support for third party API providers, and interconnection of CAPIF providers.

3GPP TS 29.222 V18.1.0 lists the service APIs in Chapter 5, and Annex A provides corresponding representations of all APIs defined in the present specification, in YAML format [4].

4.2.5.2 SEAL

The SEAL is a Release 16 development that supports vertical applications such as vehicle-to-everything (V2X), as per 3GPP TS 23.434, for application plane and signaling plane entities. It facilitates application-enabling services that are common across vertical applications. The SEAL also defines the northbound APIs for its services with vertical applications [5].

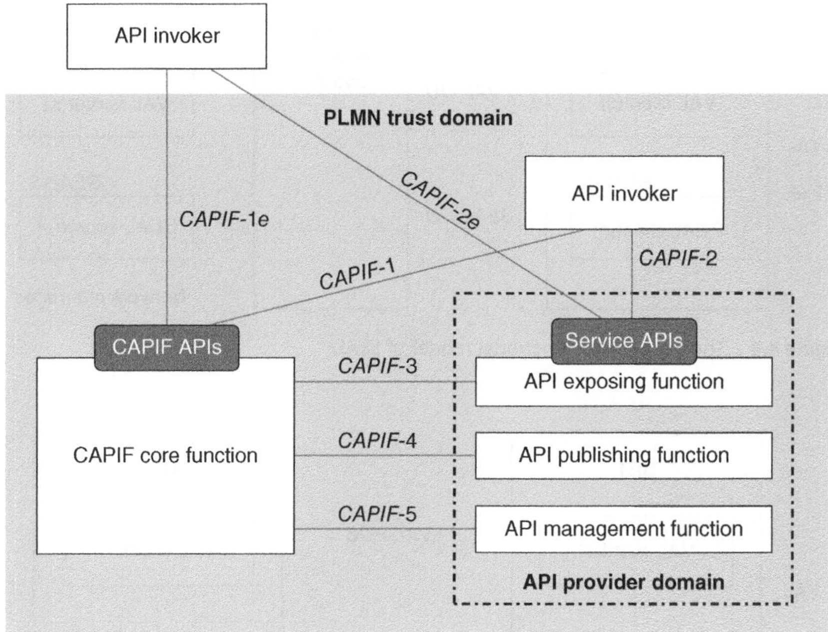

Figure 4.3 Functional model of CAPIF.

Figure 4.4 depicts the generic on-network functional model for SEAL, which can also be expanded to cover interconnection between SEAL servers, and for interservice communication between SEAL servers.

On the server and user equipment, the SEAL functional entities are grouped into SEAL clients and SEAL servers, respectively. The SEAL has a common set of services and reference points. The SEAL offers its services to the vertical application layer (VAL) [5].

Figure 4.5 depicts the off-network functional model of SEAL.

In the off-network scenario, UE1's VAL client communicates in the vertical application layer with UE2's VAL client using the VAL-PC5 reference point while UE1's SEAL client interacts with the corresponding UE2's SEAL client via SEAL-PC5 reference points. If UE1 is connected to the network via the Uu reference point, it can also act as a UE-to-network relay to facilitate UE2's access to the VAL servers via the VAL-UU reference point [5].

4.2.5.3 V2XAPP

V2XAPP refers to the application layer support for V2X services and is defined in Release 16 to enable efficient use and deployment of V2X services over 3GPP

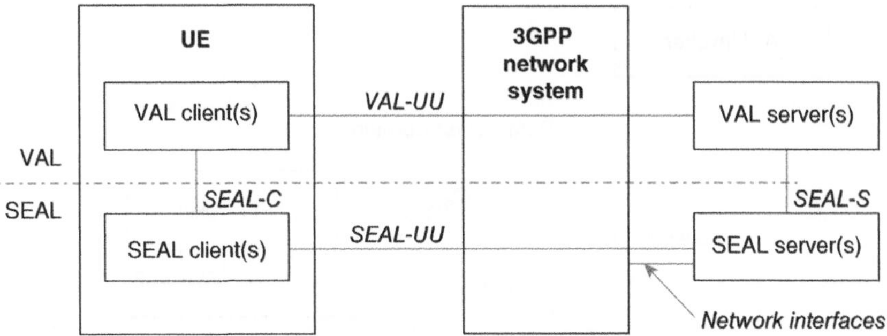

Figure 4.4 The on-network functional model of SEAL.

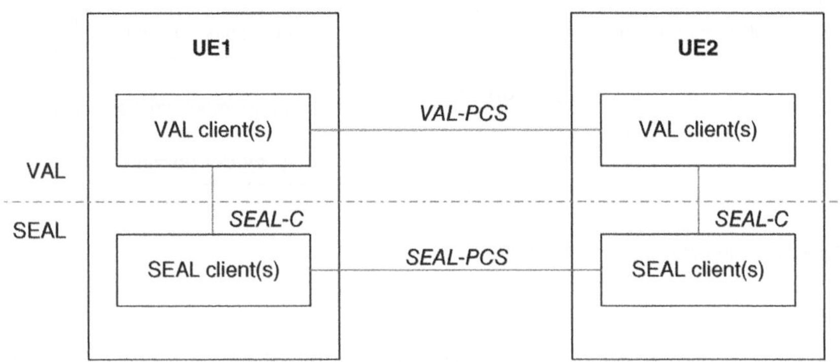

Figure 4.5 The Off-network functional model of SEAL.

systems in 4G, as per 3GPP TS 23.286 [6]. It is built on a V2X application enabler (VAE) layer that interfaces with V2X applications. The V2XAPP enables capabilities as defined in 3GPP TS 23.285, including V2X message distribution, service continuity, and application resource management [7].

Figure 4.6 depicts the V2X application layer's architectural model as per the reference model of 3GPP TS 23.285 and TS 23.287.

The V1 reference point supports the V2X application-related interactions between V2X UE and V2X application server (AS), in unicast and multicast delivery modes. The V5 reference point is for the interactions between the V2X UEs.

The V2X application layer's functional entities for the V2X UE and V2X application server are grouped into the V2X application specific layer and the VAE layer.

- The VAE layer provides VAE capabilities to the V2X application specific layer.
- The V2X application layer functional model utilizes the SEAL services as per 3GPP TS 23.434.

Figure 4.6 V2X application layer architecture.

The VAE server is in the VAE layer. The SEAL services that the VAE layer uses are location, group, configuration, identity, key, and network resource management. The V2X application server, in turn, has the VAE server, SEAL servers, and V2X application-specific server. The VAE server offers to the V2X application-specific server V2X application layer support functions via the Vs reference point, whereas the SEAL servers provide the SEAL services to the V2X application-specific server via the SEAL S reference point.

4.2.5.4 EDGEAPP

3GPP Release 17 introduced further new vertical applications such as EDGEAPP architecture for enabling edge applications in 5G (3GPP TR 23.758) and the resulting TS (23.558) describes the architecture for enabling edge applications [8].

Figure 4.7 depicts the principle of the edge computing capabilities supported by the 3GPP.

In the edge computing architecture, edge enabler layer functions, such as ECS, enable other authorized edge enabler layer functions, such as EES, to access their services, and the NFs enable authorized edge enabler layer functions (application functions, AF) to access their services. CN northbound APIs (as per 3GPP TS 23.501 and TS 23.502), are utilized by authorized edge enabler layer functions via the CAPIF core function.

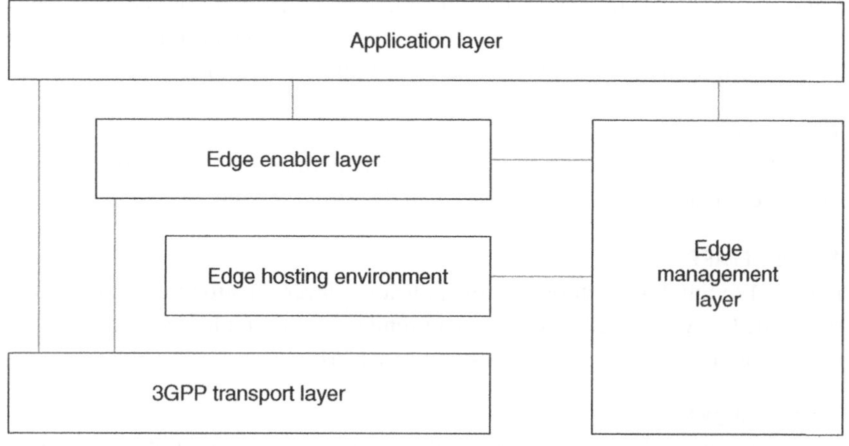

Figure 4.7 Principle of 3GPP edge computing.

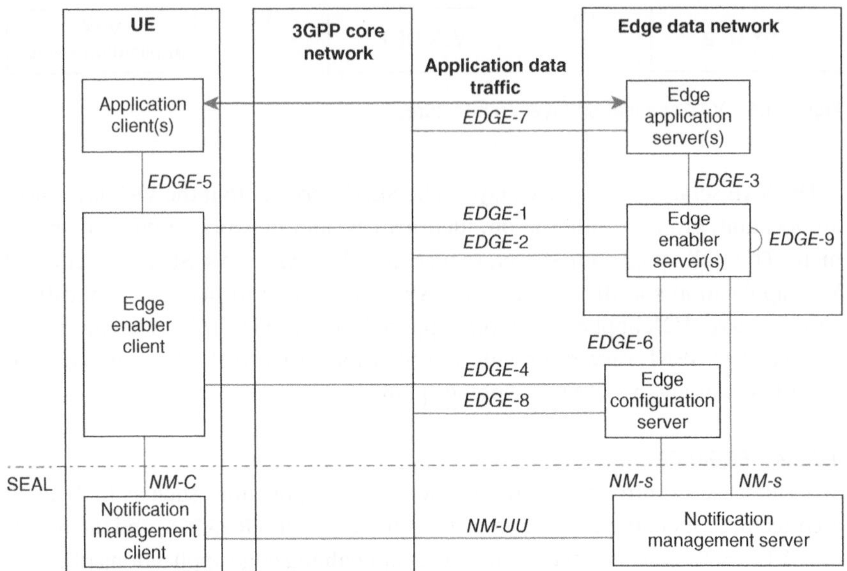

Figure 4.8 3GPP architecture for enabling edge applications.

Figure 4.8 depicts the architecture for enabling edge applications.

The architectural model shown in Figure 4.8 can also be extended to cover roaming of UEs, including both local breakout and home routed (HR) roaming scenarios. To support UEs that are roaming, the EEL uses ECSs provided in HPLMN and VPLMN.

The architectural model can also be used in federated scenarios, that is, interconnecting connecting edge entities. It should be noted that this specific item is aligned with 3GPP, CAMARA, and GSMA open platform working group reports as summarized in the "GSMA Common API Initiatives" chapter.

4.2.5.5 UASAPP

The UASAPP work item summary in Release 17 defines application layer support for uncrewed aerial system (UAS) as per 3GPP TS 22.125 and TR 23.755.

4.2.5.6 FFAPP

Release 17 includes the design of an application layer for supporting factories of the future (FFAPP) based on service requirements of cyber–physical control applications identified in 3GPP TS 22.104 and TS 22.261.

4.2.5.7 eV2XAPP

The eV2XAPP work item concerns application layer support for V2X services that enhance the capabilities of existing application layer solutions specified in V2XAPP as per 3GPP TS 23.287 [9].

4.2.6 GSMA Common API Initiatives

Telco customers benefit from common network capabilities exposed through APIs., Through its operator platform (OP) initiative, the GSMA is helping telecom ecosystems adapt a set of common APIs. The OP defines a common platform exposing operator 5G services and capabilities to customers and developers. It is a "connect once, connect to many" model. The first phase of the OP has focused on Edge, and the aim of the GSMA's operator platform group (OPG) is to expand the OP to cover other capabilities such as connectivity and network slicing. In addition to the OPG, the more concrete API development aspects are discussed in its operator platform API group (OPAG) subgroup [10].

GSMA open gateway is a concrete form of the GSMA API deployment environment, and its collaboration with Microsoft, AWS, Google Cloud, Ericsson, and 22 operator groups signed up to memorandum of understanding of interoperable APIs. The open gateway was released at the GSMA Mobile World Congress (MWC) in Barcelona in 2023. The open gateway provides federal access to network capabilities through APIs starting with the first APIs defined by CAMARA. The open gateway is intended to accelerate the commercial deployment of CAMARA APIs, and after the initial work by a corresponding GSMA Taskforce, further work is being continued by the GSMA OP and OP API working groups (OPG and OPAG, respectively). More universal APIs are expected after the initial APIs have been finalized to provide an inclusive environment for scale based on open technical and business framework templates [1].

CAMARA complements the GSMA's initiatives. Its aim is to reduce the telco network's complexity with APIs and make the APIs available across telco networks and countries for easy and seamless access. CAMARA is an open-source project initiated by the Linux Foundation to define, develop, and test the APIs, and it collaborates with the GSMA OPG to align API requirements and publish API definitions and APIs. CAMARA's key task is to harmonize APIs by creating fast and agile working code with developer-friendly documentation. Furthermore, the CAMARA API definitions and reference implementations are free to use through an Apache2.0 license [2].

4.2.7 Edge Computing

5G increasingly relies on clouds as they enable intelligent service awareness and thus optimization of connectivity, latency, and other performance characteristics. 5G networks are also highly scalable and offer advanced services due to the virtualization of the network functions. One of the consequences of this evolution is that 5G will rely largely on data centers. The 5G ecosystem has centralized, regional, and edge data centers.

In addition to centralized clouds residing in the main data centers, 5G applications may be located at the network's edge, closer to the end users. These applications can be hosted in the mobile edge computing nodes, referred to as cloudlets.

5G thus requires high-capacity cloud processing power to meet its demanding performance criteria. As 5G deployments evolve, there will be needs for higher data storage and server capacities, cooling for the equipment, and space for housing the respective racks. Often, the same data center will serve 5G and many other systems. Preparation for this can already be seen in practice as data centers are being actively deployed over wide areas.

Data centers will be changing from the previous decentralized mobile network model to improve service of the centralized processing of the 5G network functionalities. This means that instead of using stand-alone equipment for each function of the network, 5G will rely on cloud-based solutions in both main data centers and edge regions.

Thus, the 5G NFs run on virtualized software environments instead of standalone HW elements. This principle could potentially evolve further so that most of the base station processing can occur in the clouds, whereas the base station could occasionally process more, e.g. when a cloud's load reaches a predefined limit.

The edge cloud can bring both content and data processing closer to the end users. One of its clearest benefits is the reduction in latency. The closer to the mobile device the contents can be located, the lower the latency, due to the shortening of the transmission path. The edge can be located in a cloud CN or cloud RAN, or even at a base station site.

Another benefit of the cloud edge is that some of the data processing by devices such as smartphones can be offloaded to it. 5G will make this reasonable in many cases thanks to the high data speeds and low latency. Practical examples of this are uses in AR and VR devices, which typically require heavy processing power to provide fluent user experience.

"5G connectivity can be used for embedding AR/VR capacities in devices such as headsets to enhance the performance. The challenge for device's own processing of such contents is that a mobile phone cannot process data as efficiently as the standalone hardware. To overcome this restriction, the content-processing could be offloaded to edge cloud through 5G connectivity which could greatly enhance AR/VR capacities of devices such as headsets."

The increasing numbers of such demanding applications could exploit 5G connectivity and cloud computing. At the same time, the terminal cost could be reduced because it would not need so much processing power, if it supports enhanced MBB (eMBB) and ultra reliable low latency communication (URLLC) connectivity to the edge cloud.

Business-wise, edge computing and data centers open new opportunities for both established and new stakeholders. Some operators may want to deploy and manage their edge infrastructure, but it could be equally feasible to buy or lease required cloud processing and capacity based on a Cloud as a Service model.

The 5G architecture's virtualization offers completely new levels of such opportunities compared to earlier generations. The older networks are more isolated environments, where operators take care of their own stand-alone network components, i.e. machines with dedicated HW and SW running on top, whereas the functions are instance-based and HW-agnostic in 5G.

The infrastructure model may also have hybrid nature, with both own and purchased services depending on the deployment phase, area of operation, and other factors. This mode enables optimization of the return on investment (RoI), especially for smaller 5G operators, for whom building a large data center might not be rational, at least at the beginning of the deployment.

Outsourcing the cloud-based 5G NFs and other tasks generates new business models. There might also be a need to consider the assurance of service availability. Typically, an SLA is negotiated between an operator and cloud service provider to set expectations for the service up-time. In a critical environment, this may include the "five nines" criterion, i.e. guarantee that service will stay up 99.999% of the time. This may require active–active pairing of parallel servers in a georedundant configuration that provides the highest assurance, but at increased cost. For less critical applications, a cheaper passive–passive configuration without geo-redundancy may be an adequate solution.

More information on cloud computing can be obtained from ITU-T Y.3515 [11].

4.3 Vertical Aspects

4.3.1 Importance of Network Slicing for Industrial Applications

Network slicing enables operators to tailor the service level of their cellular networks so that verticals using their preferred NS can benefit from the optimal performance for their expectations. Meanwhile, the NSPs can benefit from the system's ability to allocate the invaluable network processing resources according to need. The slicing can play a key role in efficiently configuring the network to meet differing QoS requirements of diverse sets of use cases. It provides new opportunities for both established and completely new stakeholders, and both NSPs and clients can benefit from them. Many established and totally new verticals can use NSs, and they may have very different requirements. It is thus crucial for NSPs to understand their varying requirements to set the slices appropriately.

For the fluent user experiences between operators nationally and internationally, the GSMA is in a key position to facilitate deployment of adequate models through common NS templates. Hence, GSMA PRD NG.116 defines templates that operators and NSPs can consider in efforts to offer common experiences for roaming customers [12].

4.3.2 Network Slice Setup

Operators may provide network slicing based on an "as-a-service" model for the customers that enhances operational efficiencies and reduces time-to-market for new services. Network slicing is facilitated by SDN, NFV, and highly capable 5G network orchestration.

An NSP can form, manage, and terminate NSs upon need in a dynamic manner so that each slice can be optimized for a specific use case. 5G NSs can be adjusted to offer their users different user experiences within a selected area. Furthermore, the 5G UE can join one or multiple NSs, while at the same time supporting different parallel services.

The users, such as industrial applications or enterprise members, can subscribe to slices of their preference depending on availability. In practice, subscribers will see the characteristics of a set of available NSs to select those that best serve their needs. The NSP can also set different prices for NSs with different characteristics.

As outlined and exemplified by the GSMA [12] and described in 3GPP TS 28.531, NSPs can build their NS product packages based on standard and/or private slice templates (S-NESTs and/or P-NESTs, respectively). In this way, an NSP could provide a standardized network slice type (NST) "A" with a "Packet delay budget" value of 1–100 ms so that this attribute and its value range are defined by the 3GPP. The NSP could divide this definition further to offer a commercial version, including (for example) NS products "Platinum NST-A," "Gold NST-A," and "Silver NST-A" with "Packet delay budget" ranges of 1–10, 11–50 and 51–100 ms, respectively. The NSP can logically price these categories to provide the best performance for clients willing to pay a premium.

GSMA PRD NG.116 can help NSPs and other relevant organizations understand the different levels of needs and requirements of verticals in their ecosystem. It can also help establishment of a set of the most strongly desired slices within MNOs' own infrastructures and interoperable environments, including roaming scenarios that benefit from uniform user experience across the operator community. In addition, GSMA PRD NG.127 provides a relevant reference for vertical needs.

GSMA NG.116 serves as a common guideline for network slicing templates that MNOs can choose, with commonly used attributes (e.g. data speed) and values

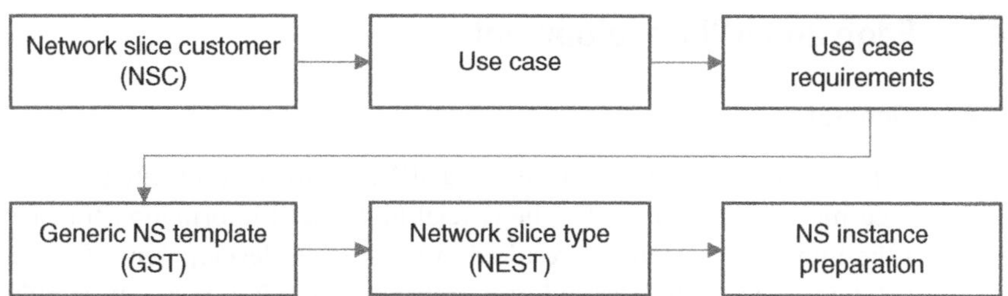

Figure 4.9 The overall process of setting up a GST and NEST [12].

(e.g. maximum, and average values in terms of Mb/s). Figure 4.9 depicts the common process for setting up a GST and NEST, based on the best understanding of the use cases and associated needs that the NS is intended to meet.

For an NSP to set up an NS adequately for its performance to meet users' expectations, it is essential to capture the users' requirements. This is not necessarily straightforward in practice as the users might not be able to clarify their real needs in technical terms. The GSMA network slicing Taskforce explored this issue and regional vertical needs of selected use cases during 2019–2021. The Taskforce then used the acquired understanding to design a process for considering needs in an ecosystem, converting them to NS requirements, and deploying corresponding NSs. The process involves the following practical steps:

- *Regional Verticals*: An NSP/MNO should first identify and assess needs of verticals that can benefit from network slicing and prioritize them in terms of factors such as timings of the needs and significance of the markets the verticals represent.
- *Vertical Needs Assessment*: The MNO and/or NSP should reach out to prioritized verticals to obtain more detailed understanding of the practical environment and needs for the communications technologies by studying relevant use cases.
- *Use Case Assessment*: Through understanding typical practical needs and challenges in their communications environment, the NP/MNO can form lists of relevant attributes and value ranges for selected use cases.
- *Network Slice Template Forming*: Finally, the NSP/MNO can map the attributes and values into network slicing templates (GST and NEST) to deploy actual NSs in the MNO infrastructure.

An assumption underlying this process is that the NSPs actively research the markets, but equally potential NS users can reach out to the NSPs to express their typical needs for communication links and have their requirements assessed and noted on the candidate NS lists.

4.4 Edge and API Development

4.4.1 General

5G is based on NFV, open-source software, and SDN. In this environment, practical and evolved means are required for the crucial high-speed information transfer between the interfaces, which has resulted in the further development of APIs. The distributed legacy networks and web application services may not comply with the new, very strict 5G networking requirements, but the API and REST concepts can help the new ecosystem to cope with the challenges associated with the end-to-end telecommunication system that integrates and converges various network types including wireless and fixed systems and is largely based on cloud infrastructure.

The APIs have a very important role in 5G. For example, the ITU Focus Group on IMT-2020 emphasizes their relevance in the 5G ecosystem as they facilitate the adaptation of applications and services to deal with the programmable network functions and help applications to communicate with each other in the highly virtualized mobile communication infrastructure [11]. They also state that IMT-2020 NOPs should expose network capabilities to third party ISPs and internet content providers (ICPs) via open APIs to allow agile service creation and flexible and efficient use of the capabilities. Furthermore, for IMT-2020, there are increased needs for service customization by the service providers, including some offering customers the possibility to customize their own services through service-related APIs to support the creation, provisioning, and management of services. Hence, APIs can be regarded as essential to interconnect systems and share data. APIs can help to reduce costs of the 5G ecosystem because API-interconnected systems can use common software functions and thus optimize software production.

Another important element of the 5G ecosystem is representational state transfer (REST): an architectural style designed for distributed hypermedia systems. REST has become important with the rising popularity of the geographic web as it is especially suitable for sharing information. RESTful Web services integrate into the web as transport media. They have less strict bandwidth, processing power, and memory requirements than earlier models and can communicate through firewalls and proxy web servers. REST is also highly suitable for interconnecting societies. It eases programming and collaboration between stakeholders and fosters multivendor and multi-operator ecosystems.

In a typical 5G network infrastructure, when offering services to the enterprises and end users an API gateway exposes the REST APIs to the third party applications and partners. The API gateway serves as an entry point that routes requests and has protocol conversion capacity. It is beneficial for cooperating parties such as developers who want to deploy their own user interfaces or transfer information via APIs.

Java also has essential roles in many 5G functions and procedures, as a class-based, object-oriented, largely implementation-independent programming language for generic purposes. The aim of Java is to provide means to design software once and run the Java-compatible code on different platforms without need for further adaptation.

JavaScript object notation (JSON) is a standardized, language-independent file format based on human-readable text. It can deliver data objects that have attributes with their values, and array data types. As the popularity of the XML format is declining, JSON is now a popular data format for asynchronous communications between a browser and server. It is also an adequate base for many 5G-related procedures within the 5GS and with external entities.

The mobile communication industry has also used a special version of the Java environment for SIM card production, including the Universal Integrated Circuit Card (UICC) associated Universal Subscriber Identity Module (USIM) application, and Card Operating System (COS) that is typically a card vendor's proprietary solution. Because of the variations in these artifacts, each USIM card application would need to be adjusted for all of the different operating systems, which would be a waste of time and resources. To overcome this issue, the Java Card Run Time Environment (JCRE) provides an abstraction layer between each card vendor's own OS variant and the apps running on them as Java applets. As the apps are OS-agnostic in JCRE the app must be developed only once and it is compatible with any OS which supports the abstraction. The same principle is expected to continue in the 5G era.

The Java Telephony API (JTAPI) is also relevant for the 5G era. It supports telephony call control and is an extensible API designed to be scaled and used in a range of domains from first-party call control in a consumer device to third-party call control in large distributed call centers.

4.4.2 GSMA Operator Platform

Operators are working to make the network assets and capabilities available across networks and national boundaries. For this reason, the GSMA has convened an OP group to help operators to develop and deploy a unified approach capable of coping with diverse use cases that operators are planning to address, such as those in healthcare and Industrial IoT settings. The OP is intended to be a generic platform that packages existing assets and capabilities such as voice, messaging, IP data, billing security, and identity management, as well as those that 5G enables, such as edge cloud and network slicing, thereby providing the flexibility required by enterprise users.

The goal of the unified OP is to have operator assets and capabilities consistently available across networks and national boundaries, and the associated API development is closely aligned with efforts of the CAMARA project. Phase 1 will include provision of means to federate multiple operators' edge computing infrastructure to give application providers access to a global edge cloud to run innovative, distributed, and low-latency services through a set of common APIs. and applications near their customers [13]. It will also enable means to monetize 5G capabilities, IP communications, and network slicing in a scalable and federated manner.

As depicted in Figure 4.10, federation of multiple OP instances will help application providers to reach a wide user base in a given, large area by the connected operators. The APIs will be used in the interfaces summarized in Figure 4.11.

In the OP, each operator will hold an instance independent of the deployment of others. The architecture can have a common exposure and capability framework and federation interface between operators. The OP will also allow the joining of third party enterprises.

The OP will enable operators' capabilities to be tied to network ad services functions so that they can be offered to application providers via the northbound APIs.

As depicted in Figure 4.11, the OP architecture has the following four interfaces.

- Northbound interface (NBI) for service management and fulfillment of enterprise and application provider use case requirements, in accordance with cloud

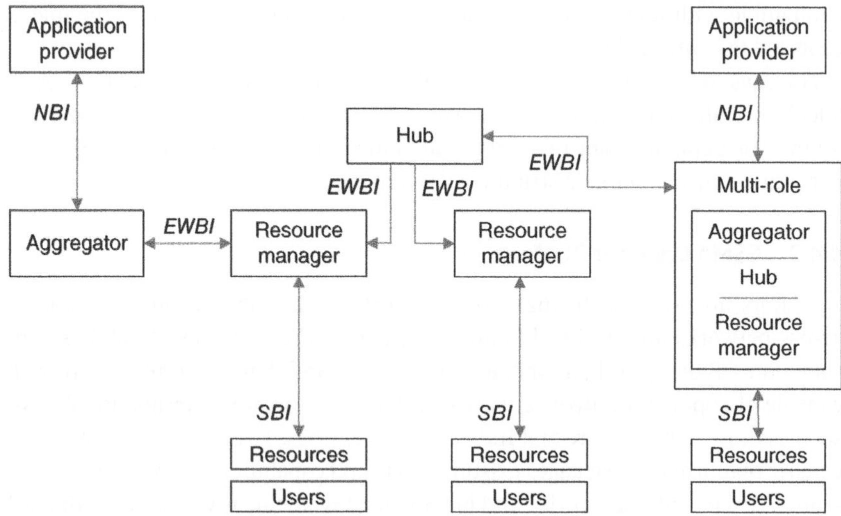

Figure 4.10 The interfaces in the federated OP, inferred from published information [13].

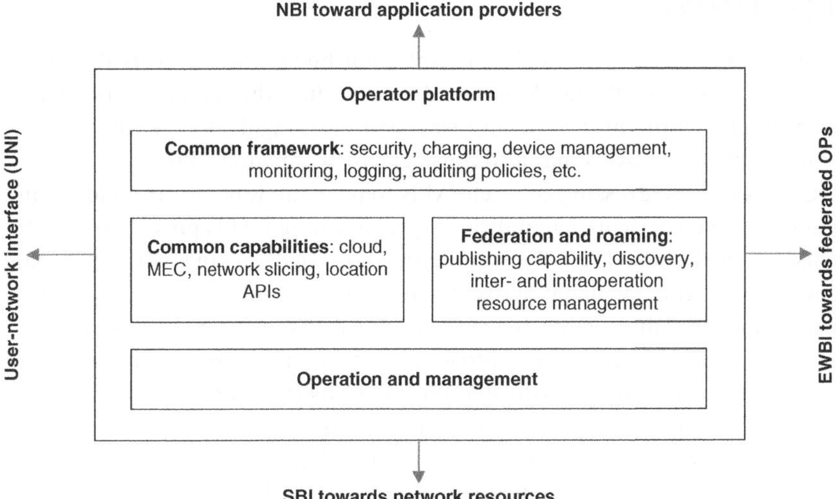

Figure 4.11 The OP interfaces, inferred from published information [13].

API principles. Network capability and service consumers could be application providers, other enterprises, or service providers
- East–westbound interface (EWBI), which extends operators' reach beyond their own footprints by their federation and allows them to exchange information.
- Southbound interface (SBI), which connects the OP with a specific operator infrastructure that delivers network services and capabilities to users.
- User-network interface (UNI), for final equipment to communicate with the OP, providing new capabilities for users, regarding (for example) dynamic service requests and location data.

The GSMA OPG has advanced further definitions of the OP, including phases and themes following the first, while the operator platform API group (OPAG) discusses the related APIs. These working groups are open to GSMA members' contributions.

As indicated by the OP themes, the APIs play an essential role in the 5G ecosystem and open opportunities for further business models and stakeholders to take advantage of the operators' network capabilities and develop evolved services benefiting enterprises and industries.

The GSMA open gateway represents the realization of the APIs. It is a framework of common network APIs designed to provide universal access to operator networks for developers [14].

4.4.3 CAMARA

A major objective of CAMARA' is to develop northbound service APIs abstracting and aggregating 3GPP and TM Forum APIs to reduce the complexity of engagement, keep control on the operator side, and meet regulatory and data privacy requirements. CAMARA may also consider east/westbound APIs.

CAMARA is addressing concrete APIs to pave the way for their more common use. Telco network capabilities exposed through APIs provide large benefits for customers [15]. By simplifying telco networking with APIs and making the APIs available across telco networks and countries, CAMARA will enable easy and seamless access. CAMARA is also considering and harmonizing service APIs, via approaches such as mapping attributes to the southbound APIs, and transformation functions (i.e. the business logic calling the southbound APIs, transforming the data, and providing the functions for the service APIs). The overall aims of CAMARA are to help telco operators to implement initial transformation functions [15] and help ecosystem members to develop APIs based on need.

Figure 4.12 presents the focal area of the telco API landscape.

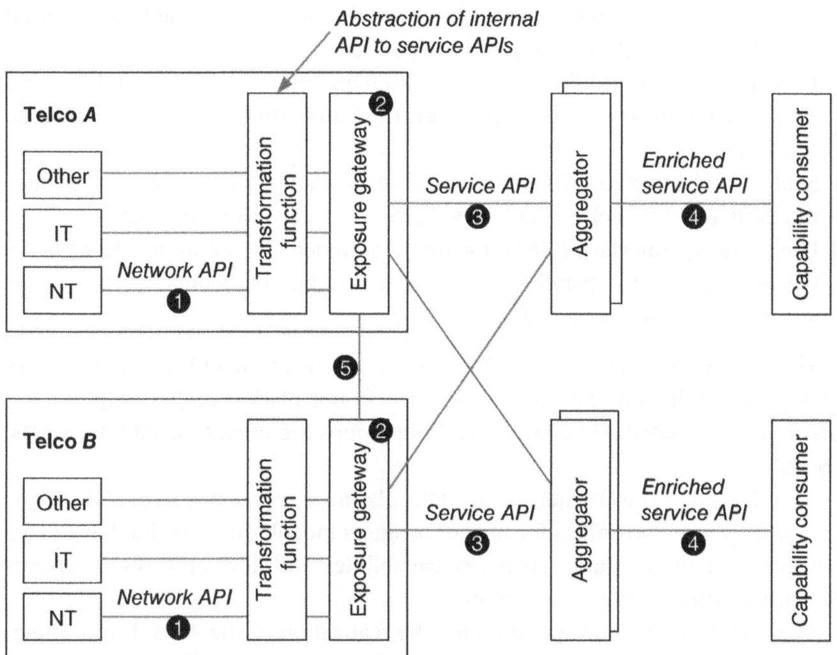

Figure 4.12 Focal area of CAMARA in the API landscape.

For the points in Figure 4.12, the following applies:

- Point (1) refers to capabilities that the network APIs expose. Examples of these APIs are related to network slicing, QoS, and positioning.
- Point (2) refers to common exposure of service functions via the service APIs across all the participating operators.
- Point (3) refers to service APIs related to, for example, billing, federation across operators, and access control.
- Point (4) refers to an optional technical aggregation that enriches service APIs involving, for example, cloud and platform providers.
- Point (5) refers to interoperability such as API roaming.

4.4.4 TM Forum

Members of the TM Forum leverage the collective intelligence of the industry through collaborative working groups of CSP and suppliers to create practical toolkits and widely adopted frameworks, including open APIs, that drive the execution of CSP digital transformation [16]. Working in open API project, its TM Forum members had collaboratively developed more than 60 REST-based open APIs by the beginning of 2023, and the industry has widely adopted them.

References

1 GSMA, "GSMA Open Gateway," GSMA, 2023. [Online]. Available: https://www.gsma.com/futurenetworks/gsma-open-gateway. [Accessed 13 March 2023].

2 CAMARA, "APIs Enabling Seamless Access to Telco Network Capabilities," CAMARA, 2023 [Online]. Available: https://camaraproject.org. [Accessed 13 March 2023].

3 3GPP, "TS 38.401, V16.2.0, NG-RAN Architecture Description," 3GPP, June 2020.

4 3GPP, "3GPP TS 29.222 V18.1.0, Common API Framework for 3GPP Northbound APIs," 3GPP, March 2023.

5 3GPP, "3GPP TS 23.434 V18.4.1, Service Enabler Architecture Layer for Verticals (SEAL), Functional Architecture and Information Flows," 3GPP, April 2023.

6 3GPP, "GPP TS 23.286 V18.1.0, Application Layer Support for Vehicle-to-Everything (V2X) Services, Functional Architecture and Information Flows," 3GPP, March 2023.

7 3GPP, "TS 23.285 V17.1.0, Architecture Enhancements for V2X Services," 3GPP, June 2022.

8 3GPP, "3GPP TS 23.558 V18.2.0, Architecture for Enabling Edge Applications," 3GPP, March 2023.

9 3GPP, "3GPP SA6 Accelerates Work on New Verticals!," 3GPP, 7 June 2019. [Online]. Available: https://www.3gpp.org/news-events/3gpp-news/sa6-verticals. [Accessed 13 March 2023].

10 GSMA, "Operator Platform Group," GSMA, 2023. [Online]. Available: https://www.gsma.com/futurenetworks/5g-operator-platform. [Accessed 13 March 2023].

11 ITU, "Y.3515: Cloud Computing – Functional Architecture of Network as a Service," ITU, July 2017.

12 GSMA, "Network Slicing: North America's Perspective, V. 1.0," GSMA, 3 August 2021. [Online]. Available: https://www.gsma.com/newsroom/wp-content/uploads//NG.130-White-Paper-Network-Slicing-NA-Perspective-2.pdf. [Accessed 13 March 2023].

13 GSMA, "Operator Platform Concept 1st Phase: Edge Cloud Computing," GSMA, January 2020. [Online]. Available: https://www.gsma.com/futurenetworks/wp-content/uploads/2020/02/GSMA_FutureNetworksProgramme_OperatorPlatformConcept_Whitepaper.pdf. [Accessed 26 April 2023].

14 GSMA, "GSMA Open Gateway," GSMA, 2023. [Online]. Available: https://www.gsma.com/futurenetworks/gsma-open-gateway/. [Accessed 26 April 2023].

15 3GPP, "TS 23.501. System Architecture for the 5G System, Release 15, V. 15.1.0," 3GPP, 2018.

16 tmforum, "About Us," tmforum, 2023. [Online]. Available: https://www.tmforum.org/about-tm-forum/. [Accessed 26 April 2023].

5

Mobile Edge and Real-Time Data-Driven Innovations

5.1 Introduction

Industrial 5G is designed as a cloud-native solution that can utilize a distributed computing and service-based architecture, enabling its applications to exploit the benefits of scale, elasticity, resiliency, and flexibility. All 5G functions and interactions, including authentication, security, session management, and aggregation of traffic from end devices, rely on modern cloud architectures – similar to those information technology (IT) businesses have been using in delivering IaaS, PaaS, and SaaS services on top of the internet for many years. A major difference, however, is that 5G enables computing to take place close to the users at the mobile edge, isolated from the internet. This new approach is called mobile edge computing (MEC), which provides users with faster response times and reduced network latency.

The more traditional mobile cloud computing (MCC) and MEC are different approaches for providing computing services to mobile users. MCC works by running applications and services in a public cloud, providing increased scalability and cost synergies by pooling a massive number of users and services into bigger entities. However, it does not provide similar service levels for the network latency and data security compared to MEC. As our industry's use studies show, the need for real-time data-driven innovations is becoming increasingly important in today's fast-paced digital era.

Here, we consider data-driven innovations as improved industrial processes that leverage real-time data and digital technologies, such as artificial intelligence (AI), machine learning (ML), and cloud computing, to improve process performance or deliver unique customer value. This new capability can lead to better decision-making, enhanced customer experience, and improved agility. Overall,

5G Innovations for Industry Transformation: Data-Driven Use Cases, First Edition.
Jari Collin, Jarkko Pellikka, and Jyrki T.J. Penttinen.

Published 2024 by John Wiley & Sons, Inc.

real-time data-driven innovation can help companies stay competitive, differentiate themselves from their competitors, and improve customer satisfaction by anticipating their needs and delivering the best possible experience. In conclusion, real-time data-driven innovation offers significant benefits for businesses and organizations across all industries.

5.2 Mobile Edge Computing and Industrial 5G

One essential capability of 5G is edge computing. The Internet of Things (IoT) is triggering an unprecedented surge in data generation, which could redefine how industries will use and process data. We have noted that Industry 4.0 and Industrial IoT (IIoT) integrate multiple types of industrial machines, devices, and assets through the network enabling data collection, analytics, and process optimization to improve productivity and sustainability. Therefore, the introduction of edge computing in IIoT can significantly reduce the decision-making latency, save bandwidth resources, and, to some extent, protect privacy. Edge computing expands the current cloud computing from cloud services closer to the end users [1].

Edge computing provides computing platforms that provide computing, storage, and networking resources, which are usually located at the edge of networks [2]. The devices that provide services for end devices are referred to as edge servers, which could be IoT gateways, routers, and microdata centers at mobile network base stations, on vehicles, and in other places. End devices, such as smartphones, IoT devices, and embedded devices that request services from edge servers, are called edge devices. With the fast development of end devices, the capabilities of computing and energy control have been significantly improved, which makes it possible to provide networking and lightweight computing services.

5.2.1 Industry Challenge, Competition, and Market Opportunity

Edge computing shrinks bandwidth costs related to long-distance data transmission and addresses the issue of processing real-time applications at the edge. Global spending on edge computing reached US$176 billion in 2022, rising 14.8% through 2021 [3]. Spending by enterprise and service providers on hardware, software, and edge solution services will continue to sustain an upward trajectory through 2025, when the figure will reach almost US$274 billion [3]. The key edge computing market drivers are as follows:

1) New digital services and applications require low latency. This dictates that the analytics assets be as local as possible to offset the latency inherent in data transmission over distance.

2) The cost of bandwidth is prohibitive for centralized computing. IoT applications generate large volumes of relatively low-value time series data, which could lead to high bandwidth charges if all of it were sent to the cloud over often costly wide area network (WAN) links.
3) There is no bandwidth available for backhaul. IoT encompasses moving assets such as freight trains and offshore oil rigs, which lack dedicated, high-speed connectivity.
4) There are security and security concerns. Although cloud providers have developed excellent security for their IoT offerings, there is a healthy amount of mistrust on the part of operation technology professionals that their sensitive data will stay secure once it leaves the walls of their organization. In addition, security extends beyond just operations and production professionals, as nation-states and other governmental bodies are reticent to share sensitive IoT data outside of sovereign boundaries, leading to the need for edge computing and near-edge aggregation.

Energy efficiency is also a driver. It has been estimated that in selected use cases, energy efficiency can be improved by 20–55% and help realize a major reduction on CO_2 emissions.

Figure 5.1 shows the market trends driving Industry 4.0 and industry digitalization. In order to respond to their customers' needs and deal with a massive data increase in the networks, all key players in the ecosystem need to invest more in their capabilities in order to modernize their digital capabilities and business processes, and improve operational flexibility and energy efficiency. Therefore, adopting edge computing as part of industrial 5G is seen to be one option to meet these requirements and to create new sources of revenue.

Since the volume of data being produced is rapidly increasing, organizations must improve scalability by distributing computing power to the edge. This reduces bandwidth costs and strain on networks, connections, and core data centers. Increasingly, the capability of analyzing data close to where it is generated is essential to responding to users' needs.

The future edge structure, including intelligent and automated orchestration across future networks and digital infrastructure (devices – Edge-Cloud), is vital in creating a seamless bridge between connectivity and cloud. This also needs to create a more solid basis for the new digital services developed by the edge application developers enabled by Edge intelligence (AI/ML). Unrelenting data traffic growth, mobile broadband, video acceleration, data centers (DCs) from mini-DCs in Class of Service (CoS) to hyperscale DCs, more subscribers with more smartphones and tablets, emerging IoT applications, and increased demand for immersive experiences, are all causing major changes to network architecture and business models. They are driving more investments in coherent optical transport,

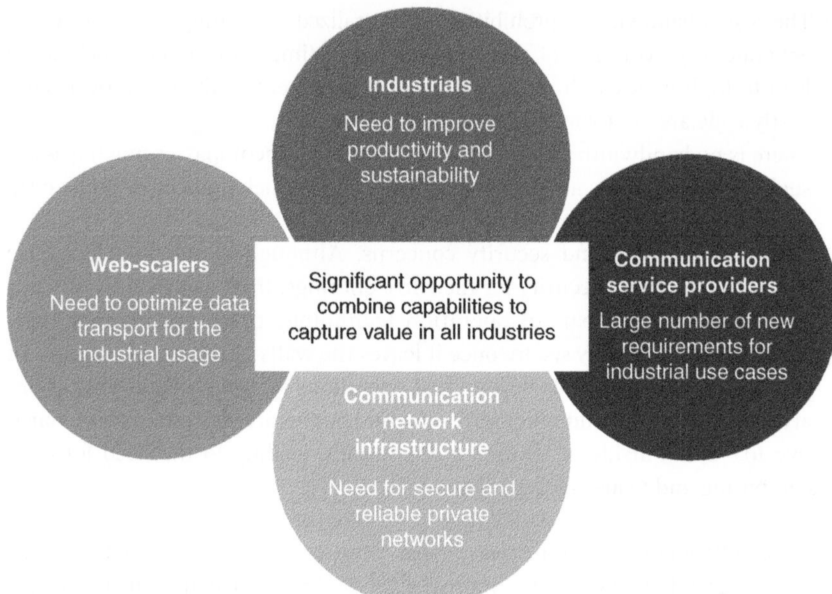

Figure 5.1 Market trends and opportunity drivers.

routing infrastructure, video caching/content delivery network (CDN), fiber access networks, distributed broadband network gateway (BNG), network control functions including policy control and enforcement, telemetry, and analytics (including AI/ML) with feedback loops for automation.

5.2.2 Operational Benefits and Use Cases Across Industries

Edge computing addresses a plethora of critical infrastructure challenges such as excessive latency, bandwidth limitations, and network congestion, leading to fast response times and improved performance [4]. Edge clouds or edge computing infrastructure will be especially beneficial in industrial environments where potentially there will be multiple IoT applications and solutions to serve a wide variety of industrial vertical-specific use cases. The main advantages of the edge computing paradigm can be summarized as three aspects [2, 4–6].

The first advantage is ultralow latency. In conventional cloud computing, all data must be uploaded to centralized servers; after computation, the results need to be sent back to the sensors and devices. This process creates great pressure on the network, specifically on data transmission costs relating to bandwidth and resources. In addition, the performance of the network will deteriorate with increasing data size. A more critical situation arises for IoT applications that are time-sensitive, meaning that very short response times are nonnegotiable. Computation usually takes place in proximity to the source data, which saves substantial amounts of data transmission time. Edge servers provide nearly real-time responses to end devices.

Saving energy for end devices represents the second major benefit. Most IoT devices have limited power (smart sensors etc.) and, to extend the lifetime of devices, it is necessary to balance power consumption by scheduling computation to devices that have higher power and computational capabilities. In addition, processing data in computation nodes with the shortest distance to the user will reduce transmission time. In a cloud computing-based service, the data transmission speed will be affected by network traffic, with heavy traffic leading to long transmission times and increasing power consumption costs. Thus, scheduling and processing allocation is a critical issue that should be considered.

The third benefit is scalability. Cloud computing is still available if there are not enough resources on edge devices or edge servers. In such a case, the cloud server would help to carry out tasks. Most IoT devices have limited computation and energy resources, with which it is impossible to undertake on-site complex computational tasks. IoT devices simply gather the data and transmit it to more powerful computing nodes, in which all the original data will be further processed and analyzed. Nonetheless, the computational capacity of individual edge nodes is limited, and thus the scalability of computational capacity for edge computing is challenging. Still, IoT devices usually do not require much computational capacity, and the demands of IoT can be properly met, especially for real-time services, by edge nodes. In addition, edge nodes mitigate the power consumption of the IoT devices through the offloading of computation tasks.

All these categories are essential across industries. For example, The World Economic Forum has estimated that digital transformation has great potential for increasing the productivity and environmental value in mining through four main digital themes: (1) automation, robotics, and operational hardware; (2) digitally enabled workforce; (3) integrated enterprise, platforms, and ecosystems; and (4) next-generation analytics and decision support [7]. The reliable high-speed connectivity provided by 5G networks and edge computing capabilities creates many opportunities for the mining industry. It provides new possibilities for operational principles and approaches for organizing mining operations, for example, by enabling new data-driven tools for managing the utilization and maintenance of machinery and other assets [8, 9]. To maximize the benefits that 5G and edge computing can provide to underground operations, use cases must be planned, designed, and deployed considering a long-term view and a holistic approach encompassing not only operational but also safety, environmental, and commercial aspects of the business. It also requires a clear definition of methodologies and critical indexes to measure and monitor the improvements provided by 5G and edge computing.

Similar use cases of edge computing can cover all industry verticals. The value proposition for IoT applications is the most noteworthy [3]. Manufacturing and heavy industries employ IoT applications for a variety of factory floor operations, not just limited to monitoring and automated operations of complex machinery. The edge, again, facilitates the integration of IoT applications for predictive analysis and maintenance. Fifth-generation mobile networks have a distributed cloud architecture based on software-defined networking principles and virtualization. They include robust IoT and device management solutions as well as analytics, ML, and AI capabilities. These capabilities are built into the network, but they can also be leveraged for specific mining applications. For instance, remote wireless automation solutions require very low-latency communications. This, in some cases, will require edge computing resources to be placed very close to the remotely controlled or automated devices. Fortunately, edge computing resources are built into the 5G architecture and can provide distributed cloud connectivity to support applications in the field, in railway tunnels, or in deep underground shafts. In retail, minimal latency could be used to create a rich, interactive shopping experience in stores or even at home. Retailers often produce a lot of data, including stock tracking, sales, and surveillance. Edge computing can help sift through the diverse data, create predictive models, and identify novel business opportunities. Healthcare is set to reap the benefits of edge computing in a major way. Data analysis from IoT devices, sensors, and other medical equipment will help swift decision-making and ensure effective medical intervention. Healthcare and hospital environments are under pressure to make real digitalization transformation. There is need for modernization of the systems and processes to increase level of automation by using edge computing to ensure realization of the benefits including low latency, data privacy, and data security. By bringing private 5G network with edge computing capabilities to a hospital environment, the level of automation in an operating theatre with secure communication and low latency can be realized in practice, for example, to enable remote surgery [10]. Autonomous vehicles will generate enormous volumes of data about vehicles, road and traffic conditions, speed, and location [11]. The information must be aggregated and studied in real time for safety and reduced journey durations. Such tasks require efficient onboard computing that can only be achieved at the edge.

Data collected from a wide range of sensors and drones from farmland could provide valuable information on soil temperature, water presence, and density of nutrients and predict optimum crop growing and harvesting periods [12]. Edge computing will boost gaming and video experiences by reducing lag time and optimizing high-definition streaming. Data from the radio network can be used to augment network performance, for instance, enabling near real-time control (RIC) for beam optimization.

5.2.3 Sustainability Benefits

Industrial 5G, including edge computing capabilities, can provide a positive contribution toward sustainable operations, mainly through the increased efficiency of operations and increased potential for innovation [5]. From a wider viewpoint, it is estimated that digital solutions could reduce global emissions by up to 20% [8]. By now, it should be clear that digitalization and sustainability are highly interconnected. For this reason, these trends are sometimes combined into the so-called twin transition. This approach recognizes that there is an untapped opportunity for technology to drive sustainability goals.

Reduction of network load can be optimized through an intelligent edge structure. This saves energy and costs as well as reduces the time between generating data and acting on them, enabling significant user and business performance benefits. Edge computing could provide a new solution for significantly saving energy on a global level. 5G and edge computing will play a significant role in promoting and attaining sustainability goals because these technologies have been designed to use energy efficiently throughout its ecosystem. The European Union (EU) aims for a 55% reduction in greenhouse gas emissions with the 2030 Climate Target Plan. For example, it has been estimated that, together, 5G and edge computing can help reduce carbon emissions from mobile networks by 50% over the next 10 years. This means that when edge-enabled use cases are implemented across all sectors in the United Kingdom, they could reduce CO_2 production by up to 269 megatons by 2035, with a carbon emission reduction from 73 million tons to 34 million tons.

Mobile networks provide flexibility for greater sustainability. According to the GSMA, mobile networks in particular provide additional flexibility in a world of connected people and things in sustainable societies and economies. In its Enablement Effect report, the GSMA estimates that while mobile networks contribute about 0.4% to global emissions, there has been a tenfold reduction avoidance in emissions and this is set to double by 2025. Mobile technologies drive two forms of enablement: using IoT to avoid emissions in buildings, transport, manufacturing, and the energy sector, and behavioral changes, for example, using more public transport and remote working.

5G drives the top five use case categories: IoT, AR/VR, autonomous driving [8], drones, and robotics. Edge computing is a key enabler for all of these. It can help reduce global energy consumption and related greenhouse gas emissions caused by data transport in the networks through data traffic optimization [9]. In addition, with well designed and optimally located edge DCs or buildings, waste heat could be more easily collected, stored, and utilized for heating the property and/or domestic water. Even if waste heat were not recovered, it would be possible to remove waste heat energy-efficiently if equipment were placed close to each other within the DCs at the design stage. This would enable an 80% CO_2 reduction when waste heat is reused.

Another example from the steel industry shows that edge computing can help drastically reduce CO_2 emissions by up to 12% by using real-time measurement data and ML models to determine the quality of the manufactured steel and ensure the product meets standards. This illustrates the wide-reaching impact across industries and domains that edge computing has.

5.3 Data Quality and Cyber Security

The quality, or rather lack of quality, of data is a big challenge for many companies. If data are not correct and trustworthy, it becomes impossible to build any industrial processes, applications, or services on it. Therefore, industry ecosystems should pay extra attention to the continuous cleanup of data and their attributes [13]. Akin to keeping production facilities and offices clean, it is mandatory to maintain data quality at a satisfactory level in different cloud environments and system databases. However, it is not appropriate to pursue perfect quality, just sufficient quality for business. Data cleaning and cleansing services become very important factors when building an industrial 5G solution for, and between, industry ecosystem players.

Generally, data quality consists of five different dimensions for any industrial internet application: (1) data accuracy, (2) data completeness, (3) data consistency, (4) data duplication, and (5) data timeliness [13]. First, the data must correctly reflect the real-world situation so that fact-based decisions can be made based on it. This dimension is called *data accuracy*. Second, the *data completeness* means that all the necessary information is available for making conclusions and no relevant data are missing. Next, for the exchange of information across organizational and company boundaries, it is extremely important that information is consistent between different systems and cloud environments. This quality dimension is *data consistency*. Fourth, *data duplication* relates to not reporting the same information more than once to prevent results from being distorted. The fifth dimension, *data timeliness*, relates to information being available quickly enough and in good time for all parties. All these five perspectives should be carefully considered when implementing new data-driven industrial processes, applications, and services. It is worth investing in the accuracy of the data in any case.

Cyber security is another important aspect to consider when designing and implementing data-driven industrial 5G solutions. Data breaches, vulnerabilities, and industrial espionage pose serious threats. The use of public cloud and industry IoT applications over the internet will significantly expand the so-called attack surface to places where there has previously been no need to consider cyber security challenges. The old methods of risk management are not enough to guarantee information security. However, the threat should not be exaggerated but taken carefully into account in the design of data-driven solutions. Nowadays, cyber security risks can be significantly mitigated using sophisticated systems, tools, and experts in the security field. Private and mobile edge cloud solutions also minimize the cyber security risks, as data is not exposed to criminals operating on the internet.

Many industrial enterprises utilize cloud-based architectures with their IT and operational technology (OT) applications, but they rely only on private cloud solutions for processing critical data. They have dedicated computing resources (servers, storage, and applications) for industry data and manage them with their own IT/OT experts to ensure full control and customization for their specific needs. Similarly, the 4G/5G private networks are often considered the safest solution to run the connections in an industry area. Although enterprises use private cloud and network solutions, the majority of applications are still run on public cloud and commercial macro networks, as illustrated in Figure 5.2.

Different applications have different levels of security, privacy, network latency, capacity, and business continuity requirements. Therefore, distinct applications are run on diverse platforms. *Mission-critical applications* that form the basis for automated industrial processes are safeguarded by limiting their access to private clouds and private networks only. Usually, critical automation systems that run and optimize the performance of production lines belong to this category. Another

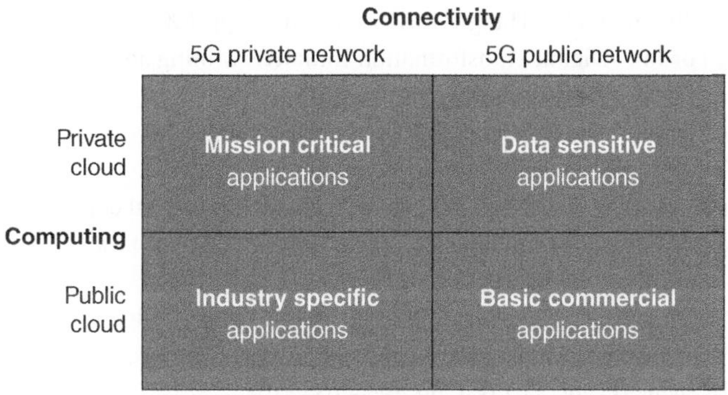

Figure 5.2 Alternatives ways of running industrial applications.

category is *data-sensitive applications*, which handle business and privacy-sensitive data and are managed only in private clouds without access to or from the internet. However, wireless connections can be based on public mobile networks, as there is no access to the internet required. Public clouds are appropriate for some noncritical *industry-specific applications,* which are run on a private network inside an industrial area. These applications are primarily designed to share data inside an industry ecosystem and are, therefore, implemented on a public cloud. Finally, *basic commercial applications* that are widely used from different locations can be safely used through public clouds and networks. It is important to manage these different access methods for each application and keep the elements of data quality in good shape.

References

1 T. Qiu, J. Chi, X. Zhou, Z. Ning, M. Atiquzzaman and D. Wu, "Edge computing in industrial internet of things: Architecture, advances and challenges," *IEEE Communications Surveys & Tutorials*, vol. 22, no. 4, pp. 2462–2488, 2020.

2 T. Tran, A. Hajisami, P. Pandey and D. Pompili, "Collaborative mobile edge computing in 5G networks: New paradigms, scenarios, and challenges," *IEEE Communications Magazine*, vol. 55, no. 4, pp. 54–61, 2017.

3 IDC, "Spending Guide Forecasts Double-Digit Growth for Investments in Edge Computing," 2022.

4 Y. Chen, N. Zhang, Z. Wu and S. Shen, Energy Efficient Computation Offloading in Mobile Edge Computing, Wireless Networks, Switzerland: Springer Nature, 2022.

5 N. Hassan, K. Yau and C. Wu, "Edge computing in 5G: A review," *IEEE Access*, vol. 7, pp. 127276–127289, 2019.

6 W. Yu, F. Liang, X. He, W. Hatcher, C. Lu, J. Lin and X. Yang, "A survey on the edge computing for the Internet of Things," *IEEE Access*, vol. 6, pp. 6900–6919, 2017.

7 World Economic Forum, "Digital Transformation Initiative: Mining and Metals Industry – White Paper," World Economic Forum, 2017.

8 World Economic Forum, "The Impact of 5G: Creating New Value Across Industries and Society – White Paper," World Economic Forum, 2020.

9 W. Dai, H. Nishi, V. Vyatkin, V. Huang, Y. Shi and X. Guan, "Industrial edge computing: Enabling embedded intelligence," *IEEE Industrial Electronics Magazine*, vol. 13, no. 4, pp. 48–56, 2019.

10 A. Abdellatif, A. Mohamed, C. Chiasserini, M. Tlili and A. Erbad, "Edge computing for smart health: Context-aware approaches, opportunities, and challenges," *IEEE Network*, vol. 33, no. 3, pp. 196–203, 2019.

11 F. S. Giust, V. Sciancalepore, D. Sabella, M. C. Filippou, S. Mangiante, W. Featherstone and D. Munaretto, "Multi-access edge computing: The driver behind the wheel of 5G-connected cars," *IEEE Communications Standards Magazine*, vol. 2, no. 3, pp. 66–73, 2018.

12 M. Uddin, M. Ayaz, A. Mansour, E. Aggoune, Z. Sharif and I. Razzak, "Cloud-connected flying edge computing for smart agriculture," *Peer-to-Peer Networking and Applications*, vol. 14, no. 6, pp. 3405–3415, 2021.

13 J. Collin and A. Saarelainen, Teollinen Internet, Helsinki: Talentum Pro, 2016.

6

Private Networks

6.1 Introduction

A non-public network (NPN), also referred to as a private network, provides network services in an isolated environment. It can be based on cellular networks or other wireless technologies, and it does not depend on numbering on regulated public land mobile networks (PLMN). 5G NPN refers to using a third-generation partnership project (3GPP) cellular system to deliver its capabilities for NPN use scenarios, e.g. businesses and municipalities. The 5G NPN can reside partially or completely within the physical premises of an organization using it, e.g. within a factory or campus area, so that an external entity separate from a mobile network operator (MNO) assumes the responsibilities for the isolated part offering its services to a limited group. NPN does not typically allow inbound roamers, although the NPN users may have roaming capabilities to use other PLMNs. The benefits of NPN deployment include the potential to control the quality of service (QoS) and protection by isolation. An NPN service can include voice connectivity in a defined geographical area, or it can focus on the Internet of Things (IoT) and respective industrial applications. Currently, there are many test projects and commercial setups involving industrial devices [1].

6.2 Standardization

6.2.1 3GPP

The 3GPP has designed 4G and 5G NPN specifications in Release 16 for enablers using Industrial 5G IoT. These definitions include time sensitive network (TNS) communications, NPN, and local area network (LAN)-type services. Release 17 evolves these aspects further. The 3GPP defines NPN in the

5G Innovations for Industry Transformation: Data-Driven Use Cases, First Edition.
Jari Collin, Jarkko Pellikka, and Jyrki T.J. Penttinen.

Published 2024 by John Wiley & Sons, Inc.

technical specifications: TS 23.251 (architecture and functional description of network sharing), TS 22.104 (service requirements for cyber–physical control applications in vertical domains), and TS 23.501 (5G system architecture).

6.2.2 Industry Bodies

The 5G-alliance for connected industries and automation (5G-ACIA) summarizes industrial IoT deployment scenarios for 5G NPNs [2] based on 3GPP-defined 5G NPNs. It presents deployment models to complement their architectural design. Also, GSMA considers NPN. As an example, their guideline [3] provides an overview of deployment of a 5G industry campus network (NPN) using the 3GPP definition, which is one of the key 5G concepts to support "to business" models (2B).

6.3 5G NPN Standard Architectures

As per the 3GPP Release 16 definitions, an NPN enables deployment of a 5G System for private use. The NPN can be deployed as a standalone NPN (SNPN) or public network integrated NPN (PNI-NPN). An NPN operator manages the SNPN without relying on the functions of a PLMN, whereas PNI-NPN deployment depends on those [4]. Figure 6.1 shows NPN variants.

6.3.1 SNPN

The SNPN uses a combined PLMN identifier (PLMN ID) and network identifier (NID). 5G user equipment (UE) supporting SNPN can be attached to it based on 5G subscriber's permanent identifier (SUPI) and credentials. The radio access network (RAN) of SNPN broadcasts the combined PLMN ID and NID in the system broadcast information and supports network selection and reselection, load and

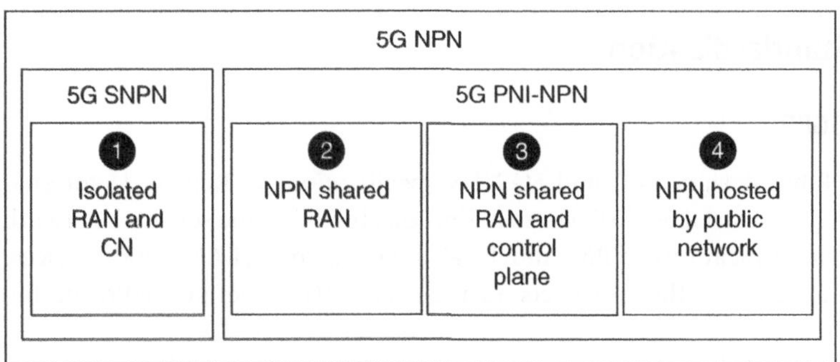

Figure 6.1 3GPP/5G-ACIA NPN variants.

access control, and barring. The NIDs can be self-assigned individually to the SNPN NIDs upon its deployment. The active NIDs may not be unique, but they use different numbering space than the other scenario, and a coordinated NID assignment that can have either (1) globally unique NID assignment independent of the respective PLMN ID or (2) a globally unique NID/PLMN ID combination.

6.3.2 PNI-NPN

The PNI-NPN uses PLMN ID whereas a closed group access identity (CAG ID) indicates CAG-enabled 5G radio cells. Within a PLMN, a CAG cell can broadcast one or more CAG IDs, in which case PLMN ID is the base for the network selection and reselection. The network uses CAG ID for the cell selection and reselection, as well as for controlling only CAG-enabled UEs access to the network.

6.3.3 Implementation Aspects

As shown in Figure 6.1 these two options result in practical scenarios of isolated deployment of a SNPN (1) or NPN in conjunction with public networks (2). The latter covers three scenarios including the industrial and IoT environment: (a) shared radio access network (RAN), (b) shared RAN and control plane, and (c) NPN hosted by the public network [5].

- *SNPN*: the NPN is separated from the public network, with all network functions (NF) residing inside the organization's premises. The communication between the NPN and the public network takes place through a firewall (e.g. N3IWF).
- *Shared RAN*: the NPN and public network share part of the RAN as per the 3GPP TS 23.251. The communication stays within NPN.
- *Shared RAN and Control Plane*: the NPN and the public network share the RAN for the defined premises while the public network carries out the network control tasks, with the NPN traffic remaining within the premises. Network slicing or 3GPP access point name (APN) can realize this case.
- *PLMN-hosted NPN*: the enterprise is served by network slicing without the need for its own infrastructure.

6.4 NPN Deployment Models

6.4.1 Standalone NPN

Figure 6.2 shows the principle of SNPN as interpreted from ref. [5, 6]. In this model, the NFs reside within the operational area of the related entity such as a factory, the SNPN being an isolated network from the PLMN. This allows communication between the PLMN and NPN through an optional firewall which

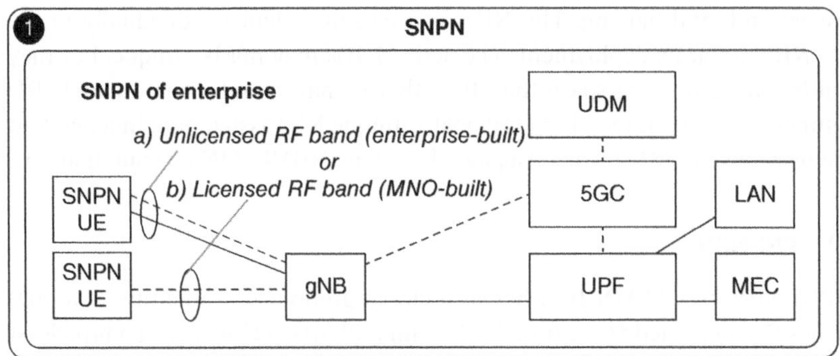

Figure 6.2 The principle of a Standalone NPN.

isolates the NPN so that an operational technology (OT) company can operate the NPN and its services, including the NPN IDs, to provide additional PLMN services in the NPN coverage area, to allow NPN subscribers to roam the public networks, and to allow public network subscribers to roam the NPN depending on the roaming agreement. NPN users may also have a dual subscription for PLMN use.

An example of the NPN 5G deployment on IIoT scenarios is the interconnection with time-sensitive networking (TSN) as per 3GPP TS 24.519. TSN is a set of new open standards that provide deterministic, reliable, high-bandwidth, and low-latency communication [7].

The 5G core (5GC) houses the NFs including unified data management (UDM) for user credentials. The user plane function (UPF) works as a data router to connect a multi-access Edge Cloud and, potentially, a local area network. The RAN manages the connectivity of 5G base stations (gNB, next-generation Node B) and UE of the SNPN users on licensed or unlicensed bands.

The standalone private network can be built in various ways: with a dedicated SLA, a local PLMN, a PLMN using a dedicated proportion of operator's licensed spectrum, or with an SA private network using unlicensed or private spectrum.

Each deployment model has their pros and cons. As an example, a licensed spectrum is one of the most expensive single items in the commercial network. A practical way to set up this type of network is to construct mm-wave radio access points in a limited enterprise area and virtualized cloud core functions in the nearby edge, with one option being a broker managing the NPN [8].

6.4.2 Shared RAN

Shared RAN involves NPN and PLMN using part of their RAN for joint use with other NF remaining separated. Figure 6.3 shows the principle of this model, depicting the connectivity of the NPN RAN to the PLMN core while the core network of

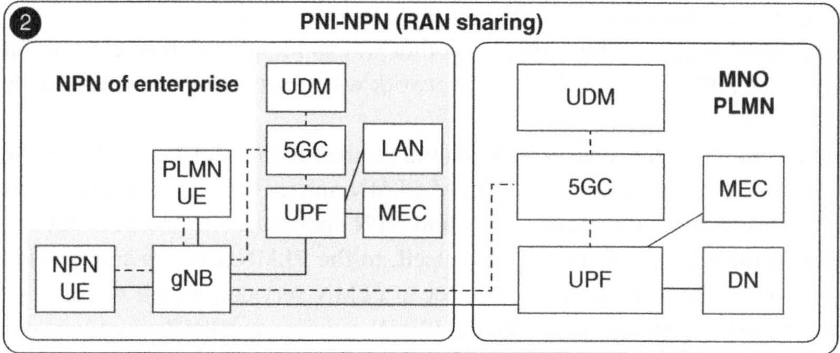

Figure 6.3 Shared RAN NPN deployment.

the NPN is isolated from the external world. In this deployment model, NPN traffic stays internal and within the logical, defined area such as factory premises.

The 3GPP TS 23.251 details the network sharing model and its architectural and functional description [9], along with scenarios for network sharing also usable in a NPN environment, such as gateway core network (GWCN) and multi-operator core network (MOCN).

The grey area indicates the private network slice that forms the NPN.

6.4.3 Shared RAN and Control Plane

In this deployment, both NPN and PLMN share the RAN within a defined business area premises, with the PLMN taking care of the control plane so that the internal NPN traffic always stays within the logical network related to the business. Network slicing is one way to set up this scenario as it creates logically independent networks within a shared physical infrastructure. The isolation of the private network portion is possible by using unique network slice (NS) identifiers (Figure 6.4).

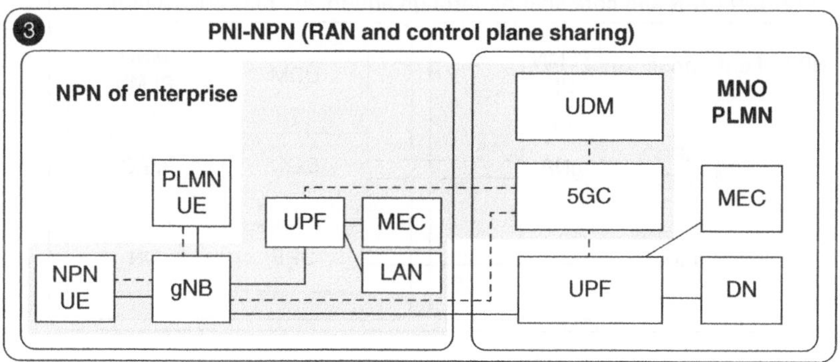

Figure 6.4 Deployment for shared RAN with control plane.

The grey area indicates the private NS that forms the NPN.

Using APN, as defined by 3GPP, is another way to implement this scenario. In this case, the APN indicates the target network with the potential to differentiate traffic.

In the shared RAN scenario with shared control plane, the PLMN hosts the NPN so that the devices are a subset of PLMN subscribers. This arrangement eases the contractual aspects of PLMN and NPN operators, and the NPN devices can also connect, apart from the NPN itself, to the PLMN services and provide roaming. The NPN services may connect to PLMN services, which requires an optional interface between the NPN and PLMN services, so NPN devices can connect to NPN services through the PLMN if the device is located outside of the NPN coverage and still within the PLMN. Logically, if the NPN devices can access the PLMN services, this interface is not needed.

6.4.4 PLMN-Hosted NPN

In this case, as a result of the network virtualization and cloudification, both the PLMN and NPN traffic are external to the business area so these traffic flows are served by different networks, and the NPN subscribers are, in fact, public network subscribers. The NPN-PLMN roaming implementation is straightforward as the traffic routes through the PLMN (Figure 6.5).

The grey area indicates the private network slice that forms the NPN.

6.4.5 Private Network on Network Slice

Although the differentiation between certain user types is possible in 4G networks, it is limited to techniques such as the isolation of services in a common infrastructure. The methods for this include APN routing, MOCN, and dedicated

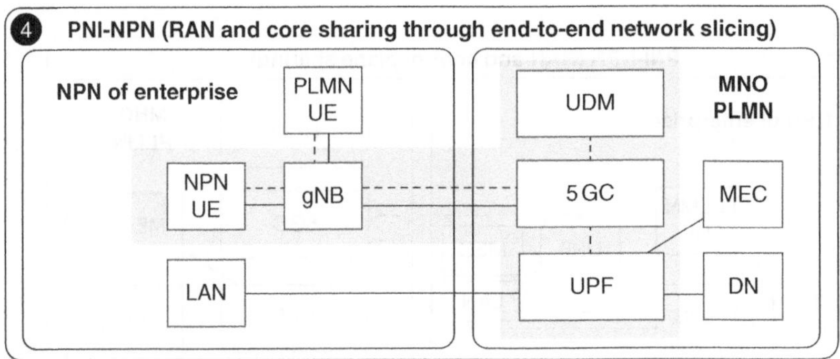

Figure 6.5 PLMN-hosted NPN.

core network (DECOR) [10]. Built upon service-based architecture (SBA), which enables the use of common hardware that executes NF as instances, 5G has been designed also to support network slicing with respective QoS assurance. In this manner, the network slice provider (NSP) can offer suitable characteristics within their different NSs, fulfilling a variety of different requirements for their subscribed verticals, also in an NPN environment. NS provides the means to differentiate the network resources and performance figures that can also create new business models. As an example, the operator can offer their customers gold, silver, and bronze categories, each having personalized NS prices and QoS figures [11]. The NSP can be either an MNO or third party. Technically, the NSO could even be used for enterprise networking if their skillset includes managing slices.

In an NS-based NPN, it is important to interpret the end-user requirements adequately. GSMA provides guidelines for NS setup based on the requirements [12] and clarifies how such requirements can be interpreted from the vertical field perspective [11].

NSs are not yet used widely in commercial standalone (SA) 5G networks, despite a forecast indicating about a 25% user base in 5G by 2025 [13–15]. Furthermore, the optimal NS functionality in practice, especially in the end-to-end scenarios, might require further development to cope with the impacts of real-world non-idealities in synchronization, near real-time orchestration, and overall management of the slices.

6.4.6 Open RAN as a Private Network

Open RAN refers to the overall movement of the telecom industry to disaggregate hardware and software to create respective open interfaces in between [16]. The O-RAN alliance publishes RAN specifications, releases open software for the RAN, and supports O-RAN alliance members in integration and implementation testing. It works on open, interoperable interfaces, RAN virtualization, and big data-enabled RAN intelligence [16, 17]. Open RAN may offer the practical potential to form small-scale, isolated shared RAN networks also in an NPN form. The concept is still evolving though, and may not provide optimal techno-economic solutions soon; nevertheless, as the technology matures, Open RAN may offer competitive 4G and 5G NPN variants.

6.4.7 Legacy Networks as a Service

Increasing numbers of incumbent MNOs have already switched off their 2G and/or 3G legacy mobile communications networks or are aiming to sunset them soon. Nevertheless, there will exist scenarios involving IoT devices served by legacy systems, especially in developing markets. According to GSMA statistics [15], the 2G and 3G systems still represent more than 20% of the global footprint in 2025. This can be a niche opportunity to manage part of this diminishing infrastructure on the remaining spectrum, maintaining a minimum feasible infrastructure for IoT private vertical networks needing only low capacity.

6.4.8 Local and Fixed Wireless Access (FWA)

The LTE-WLAN aggregation (LWA) as per 3GPP Release 13 allows mobile device configuring on simultaneous LTE and Wi-Fi links [18]. 5G can also work in parallel with non-3GPP accesses to comply with lightweight private networks' needs. 3GPP Release 15 defines non-3GPP interworking function (N3IWF) allowing Wi-Fi access points delivered using 5G infrastructure as Wi-Fi hotspots or FWA. The Release 16 defines further residential gateways (RG) to interconnect end-users' devices through trusted access points. These gateways complement the N3IWF.

FWA can provide a solution to a variety of use cases such as tethering and mobile broadband (MBB), best-effort FWA, and speed-based QoS [19]. Reflecting these use cases, the FWA could serve as a small-scale home office solution too, with quick and economic deployment without the need for fiber optics.

6.4.9 Non-3GPP Wireless Private Network

Technically, it is possible to set up a simple private network using any wireless access beyond the 3GPP specifications on an unlicensed, shared spectrum. An example of this is a Wi-Fi hotspot network using common communications applications within the network (over-the-top [OTT] voice and messaging applications). Also, low power wide area networks (LPWAN) can provide a feasible communications channel for Industrial Internet of Things (IIoT).

6.5 Summary

An NPN can serve many verticals and their use cases in a more optimal way than a public mobile communications network can. From the end-users' perspective, it is important to have enough data to understand the cost-efficient deployment models that meet the requirements, as the assessment of the scenarios prior to decisions concerning the most feasible deployment model can have a significant impact on the business. It can be assumed that 5G, along with its capability to satisfy vertical needs in a highly dynamic manner and the ease with which a hosting entity can configure the service, may turn out to be one of the most appealing solutions for enterprises and industries alike.

References

1 GSMA, "Internet of Things," GSMA, 2022. [Online]. Available: https://www.gsma.com/iot/manufacturing/private-networks/private-networks/. [Accessed 18 December 2022].

2 5G-ACIA, "5G Non-Public Networks for Industrial Scenarios," 5G Alliance for Connected Industries and Automation (5G-ACIA), 2020.

3 GSMA, "5G industry campus network deployment guideline," 10 November 2020. [Online]. Available: https://www.gsma.com/newsroom/wp-content/uploads//NG.123-v1.0.pdf. [Accessed 18 December 2022].

4 D. Chandramouli, "5G for Industry 4.0," 13 May 2020. [Online]. Available: https://www.3gpp.org/news-events/3gpp-news/tsn-v-lan. [Accessed 2 October 2023].

5 5GACIA, "5G Non-Public Networks for Industrial Scenarios," 5GACIA, July 2019.

6 H. J. Son, "7 Deployment Scenarios of Private 5G Networks," NetManias, 21 October 2019. [Online]. Available: https://www.netmanias.com/en/?m=view&id=blog&no=14500. [Accessed 22 January 2023].

7 5G-ACIA, "Integration of 5G with Time-Sensitive Networking for Industrial Communications," 5G-ACIA, 2021. [Online]. Available: https://5g-acia.org/whitepapers/integration-of-5g-with-time-sensitive-networking-for-industrial-communications/. [Accessed 18 December 2022].

8 Syniverse, "Managing the Complexities of Connectivity for Enterprises and Operators," Syniverse, 24 March 2022. [Online]. Available: https://www.syniverse.com/blog/connectivity/managing-the-complexities-of-connectivity-for-enterprises-and-operators/. [Accessed 24 January 2023].

9 3GPP, "3GPP TS 23.251 Network Sharing; Architecture and Functional Description (Release 17)," 3GPP, March 2022.

10 Samsung, "Technical Whitepaper: Network Slicing," Samsung, 2020.

11 GSMA, "Network Slicing: North America's Perspective V1.0," GSMA, 3 August 2021. [Online]. Available: https://www.gsma.com/newsroom/wp-content/uploads//NG.130-White-Paper-Network-Slicing-NA-Perspective-1.pdf. [Accessed 18 December 2022].

12 GSMA, "Generic Network Slice Template V 7.0," GSMA, 17 June 2022. [Online]. Available: https://www.gsma.com/newsroom/wp-content/uploads//NG.116-v7.0.pdf. [Accessed 18 December 2022].

13 GSMA, "5G Global Launches & Statistics," GSMA, 2023. [Online]. Available: https://www.gsma.com/futurenetworks/ip_services/understanding-5g/5g-innovation/. [Accessed 11 January 2023].

14 GSA, "5G Market Snapshot 2021 – End of Year," GSA, 2023. [Online]. Available: https://gsacom.com/paper/5g-market-snapshot-2021-end-of-year/. [Accessed 11 January 2023].

15 GSMA, "Mobile Economy Report 2022," GSMA, 2022. [Online]. Available: https://www.gsma.com/mobileeconomy/wp-content/uploads/2022/02/280222-The-Mobile-Economy-2022.pdf. [Accessed 11 January 2023].

16 M. Shelton, "Open RAN Terminology," Parallel Wireless, 20 April 2020. [Online]. Available: https://www.parallelwireless.com/open-ran-terminology-understanding-the-difference-between-open-ran-openran-oran-and-more/. [Accessed 29 July 2020].

17 O-RAN Alliance, "O-RAN: Towards an Open and Smart RAN," O-RAN Alliance, October 2018.

18 Intel, "LTE-WLAN Aggregation (LWA): Benefits and Deployment Considerations," Intel, 2016. [Online]. Available: https://www.intel.com/content/dam/www/public/us/en/documents/white-papers/lte-wlan-aggregation-deployment-paper.pdf. [Accessed 8 January 2023].

19 Ericsson, "Fixed Wireless Access Handbook 2023 Edition," 2022. [Online]. Available: https://www.ericsson.com/en/fixed-wireless-access. [Accessed 8 January 2023].

Part II

Industry Case Studies

7

Mining Industry: Striving for Autonomous Connected Operations Underground

CASE STUDY TEAM MEMBERS:

Jagdeesh Rajani: Aalto University
Miika Kaski: Sandvik
Teemu Härkönen: Sandvik
Ville Svensberg: Sandvik
Jyrki Salmi: Oulu University
Jari Collin: Aalto University

7.1 Introduction

Sandvik, the mining industry case company, is a global, high-tech engineering group providing solutions that enhance productivity, profitability, and sustainability for the manufacturing, mining, and infrastructure industries. In 2022, the group had approximately 40,000 employees, sales in around 150 countries, and revenues of about SEK 112 billion from continuing operations. Their offering covers the entire customer value chain and is based on extensive investments in research and development, customer insights, and deep knowledge of industrial processes and digital solutions. The company has a world-leading position in equipment and tools, service, digital solutions, and sustainability-driving technologies for the mining and infrastructure industries, such as automated and electric mining equipment and eco-efficient rock processing.

The mining industry represents a major customer segment that includes large underground and surface mines where sustainability, safety, and productivity with automation play a central role. As a global OEM (original equipment manufacturer) supplier, the case company provides mines with advanced automation

5G Innovations for Industry Transformation: Data-Driven Use Cases, First Edition.
Jari Collin, Jarkko Pellikka, and Jyrki T.J. Penttinen.

Published 2024 by John Wiley & Sons, Inc.

and teleoperation systems, significantly improving safety and productivity while lowering the total cost of ownership. The offering consists of mining equipment, mining machines, and rock excavation, which cover a variety of products, e.g. rock drilling, rock cutting, crushing, screening, loading, hauling, tunneling, quarrying, and breaking and demolition. In addition to the core products, services have an increasing role in the business, e.g. digital service solutions, maintenance programs, inspections, rebuild solutions, safety solutions, and financial services.

Customer focus and innovation are the company's core values that boost the digitalization of mining technology for the benefit of its customers. Helping customers to make the most of digitalization to improve productivity and become more sustainable is important, from connected mines with self-driving machines and solutions for predicted maintenance to closed-loop production solutions and digital twin technologies. Using the company's products and services, customers globally can accelerate the digitalization of mining processes, which creates a huge opportunity for them to automate operations and maintenance as well as to improve sustainability and safety. Digitalized operations require modern wireless communication and computing technology solutions.

In recent years, the rapid development of information technology (IT), communications technology (CT), and operational technology (OT) has affected nearly every facet of the mining process. Mining companies are globally deploying these new tools and applications to reap the associated productivity and financial benefits. However, they face a key challenge in that they require the appropriate infrastructure to support data CT in the mining environment, particularly underground mines [1].

Reliable communication is – and has always been – a mandatory part of safe operations in a mine area. Today, local communication systems are operated with multiple, parallel fixed and wireless communication systems that ensure redundancy and applicability with both legacy and modern mining technologies. Wireless local area network technology (WLAN/Wi-Fi) has already been used in mines for years – both aboveground and underground – using unlicensed radio frequency spectra. It flexibly enables the extension of the local IT network and provides users with limited mobile and basic location-based services. There are also some private LTE (4G) networks in use, but the vast majority of wireless mining networks are still operated using Wi-Fi technology.

Automated mining operations based on Wi-Fi technology, however, include many pitfalls and challenges, as the CT is not primarily designed for running industrial processes. First, building sufficient network coverage and capacity underground is very complex and labor-intensive. Radio network planning and design require extra effort to guarantee full line-of-sight, which is needed by Wi-Fi technology, in all corners of the tunnels. This eventually may result in a huge number of base stations that need to be operated and maintained underground.

Second, connection reliability does not meet all standards of industrial processes, e.g. hand-over between two Wi-Fi base stations is not seamless and fast enough. For instance, when vehicles, machines, or devices are moving from one base station area to another, industrial processes may be disturbed or even stopped. Third, the latency of data communication and its standard deviation do not always meet the requirements that are needed to handle real-time data for automated operations effectively. Too high a latency with a fluctuating standard deviation can cause various problems to automated processes in the mine area. The fourth pitfall is linked to the cyber security requirements of industrial operations, which are often extremely rigorous. When using a local Wi-Fi network for industrial purposes, the customer is ultimately responsible for the configuration of terminal ID identification, authentication, and data encryption levels. This requires good expertise in network and IT security domains, which is often not the case in traditional mines.

7.2 Industry Transformation Challenge

Industry digitalization is advancing quickly in mines across the world. Sandvik's ambition is to develop globally scalable solutions for the mining industry by utilizing digitalization in close cooperation with its industrial partners, research institutes, and universities. There are four key pillars for digitalization: autonomy, wireless, environmental sensing, and data analysis with machine learning. The autonomy of mining operations, functions, and tasks planned is at the center of digitalization. It also takes an up-to-date environment, other nearby machines, and a given freedom of autonomous planning into account. Reliable wireless connectivity (voice and data) is another mandatory pillar needed to enable smart and autonomous mobile machinery systems in the mines. Here, connecting and collecting sensor data in real-time from machines, vehicles, devices, people, and processes is a required capability. The third area is spatial sensing, which requires a holistic environmental awareness with sensor fusion across wide areas at a mine site. The last pillar is based on data analytics and machine learning that enable autonomous machines to learn continuously to survive in more complex operating scenarios.

As a forerunner in mining technology, the case company has invested in the latest wireless communications solutions to advance digital transformation in the industry. The company has deployed a 5G standalone (SA) private network at its test mine to test, develop, and prototype mining solutions for its customers worldwide. The network enables fast, reliable, and secure voice and video communications in a real-life mining setting underground. Its 5G capability is also used for automated mining processes, enabling remote machine operations over

4K video links between areas deep underground and the surface control center. These capabilities open new opportunities in robotics, remote and autonomous operations, full-fleet automation, analytics, and enhanced safety. As such, the 5G private network comprises a breakthrough in the digital transformation of mining.

In addition, the test mine surface area is covered with a virtual 4G/5G private network based on slicing and edge computing capabilities on a commercial mobile network. The virtually implemented private network connects the test mine to the edge of the mobile network and enables the development of digital mining solutions with cloud-based applications utilizing secure edge computing centers located near the test mine. When creating a virtual 5G private network, a slice of the public 5G network is dedicated to the critical connectivity needs of the test mine. The solution is scalable, so that the virtual network infrastructure can be used by the customer's ecosystem partners in the same limited area. Compared to previous wireless technologies, 5G is expected to provide users with speed levels tens of times higher, connection reliability with latency of a few milliseconds, hundreds of times more capacity for IoT devices per square kilometer, improved security, and several new features that are important, especially for industrial use.

Sandvik is a technology forerunner OEM company in the global mining industry for testing and implementing industrial 5G solutions for improved productivity, safety, environmental sustainability, and global competitiveness. Together with its industry ecosystem companies and research institutes, a recent research project – Next Generation Mining (NGMining) – brought together industrial 5G private networks, edge computing, and AI technology-based solutions to enable digital transformation in mining. The core theme of the research project was to facilitate safe and sustainable underground mining through productive use of autonomous and connected machinery [2]. The target of the research was to develop a mine-compliant connectivity infrastructure, with integrated solutions that incorporate safety and tracking technologies and AI enablers, for safe and efficient operation of autonomously connected working machines (AGVs) in underground mines [2].

Industrial 5G technology is expected to increase the efficiency and safety of mining operations by improving situational awareness and increasing the level of autonomy of mining machinery [3]. Leading mines are already convinced of the benefits and are massively implementing private 5G SA networks underground. According to a leading mine, the new private 5G networks improve communication, further develop operational safety, increase productivity, and maintain positive mine performance as well as enable the development of new technologies in collaboration with the partners. To support these leading customers and prepare for upcoming customer deals, the case company needs to better understand what concrete benefits and new business opportunities 5G technology enables in building, operating, and maintaining wireless solutions in a multi-vendor mining environment.

7.3 Data-Driven Use Cases

Autonomous machines and robotic systems are becoming more affordable and effective in mining. This availability of remote-controlled equipment and vehicles is becoming commonplace and will continue to provide mine operators with a safer working environment. Today, robotic machines can operate in all areas of a mine. From a mining drill providing ore for electric trains to carry it to the surface, to dump trucks following a prescribed path to the processing facility, all can be controlled with preset instructions to receive a load at point A and to dump it at point B, for example. Digital innovation in mining is continuing apace. Miners across the world have embraced digitalization, understanding that its principles and applications provide a wide array of opportunities to drive cost improvements and increase company-wide productivity [4].

7.3.1 Introduction to Use Cases

In the mining process, there are many practical use cases where sensor data from machines and vehicles are turned into meaningful information and automated operations by utilizing real-time 5G connections. AutoMine® is Sandvik's future vision and concept for autonomous mining equipment based on the latest technologies and equipped with completely new sensing capabilities and artificial intelligence to enhance mining operations. It allows worldwide customers to scale up automation with new digital capabilities at their own pace. Sandvik's Next Generation of Autonomous Drilling is the latest, most compelling data-driven use case employing 5G technology. It is a fully autonomous, twin-boom development drill rig capable of drilling without human interaction. The cabinless battery-electric drill can plan and execute the entire drilling cycle, from tramming to the face, setting up for drilling, drilling the pattern, and returning home to charge for the next cycle. The underground drill has no operator cabin, creating space for onboard water and battery storage and eliminating the need for supply cables or water hoses during operation (Figure 7.1).

The self-contained drill uses and optimizes power and electricity based on need, making the onboard supply last even longer. The drill only needs to know which tunnel and face it should tram to and can plan the rest of its mission autonomously, using data from the preferred mine planning software. Tunnel lines and profiles are planned and defined into drilling and blasting patterns in a separate IT system that ensures optimal hole placement, detonation, and profile quality. The drill also has access to three-dimensional (3D) models of a site, which are automatically merged from survey and mapping results. It can update and optimize the 3D model of the mine in real-time based on feedback from its onboard cameras and scanners. The drill learns and adapts to the ever-changing

Figure 7.1 Sandvik AutoMine® Concept Underground Drill. *Source:* Reprinted with permission from Sandvik AB.

environment to complete entire missions from entry to exit safely. It navigates tight spaces with agility, enabling the effective use of large machines in small spaces. Once it has arrived at its destination, it uses automatic drill plan adjustment to optimize the drilling pattern and drill the full round.

An automated unmanned truck, such as a dump truck or a loader, is another practical use case where 5G technology is, today, already providing clear benefits for mining processes. Intelligent mine automation system can manage dumpers in motion and optimize the process remotely. It reduces damage and repair work and adds the highest levels of efficiency to give a lower cost per ton. It is scalable for different mining applications and can be supervised from a remote location (Figure 7.2).

In addition to a single truck, this advanced automation system can be used to manage the fleet of trucks sharing the same automated production area. It provides automatic mission control and automatic traffic management for the equipment fleet, while system operators remotely supervise the process. The fleet solution is designed for mining operations, where good fleet performance is required in large production areas shared by multiple pieces of equipment. These applications include, for example, block caving environments, ramp haulage, and transfer levels. The system offers interfaces to external mining systems for various purposes, such as mine planning for accurate draw control.

Figure 7.2 Sandvik AutoMine® Loader. *Source:* Reprinted with permission from Sandvik AB.

Sandvik AutoMine® can also be utilized in a larger automation system where one system operator can remotely control and simultaneously supervise multiple automated underground loaders and trucks. Each piece of equipment completes automated missions in its dedicated production area. In addition, intelligent teleoperation with operator-assisted automatic steering is possible with this solution. The solution provides a powerful way to take advantage of the full machine performance through automation and offers substantial benefits of increased productivity, safety, and cost-efficiency in mining operations.

7.3.2 Performance Tests Between 5G and Wi-Fi

An unmanned and remotely controlled loader was selected as a practical use case to compare the performance between the private 5G SA and Wi-Fi networks. The testing was carried out in real-life settings while the loader was running a series of predefined tasks underground. The performance tests were carried out in a test mine within the case company's R&D center. The measurements took place underground in a limited area inside the test mine. The test area covered a roughly 200-m-long route between the dumping site and collection site, between which the unmanned loader vehicle (like a dump truck) was driven from the control room back and forth. The loader vehicle was installed with two clients: Nokia 5G and Cisco Wi-Fi access points. A 5G network operates in the 3.5 GHz frequency band, whereas the Wi-Fi band utilizes the 2.4 GHz free spectrum band. The target was to evaluate the applicability of these two technologies based on selected KPIs: uplink (UL) throughput, latency (RTT, round trip time), jitter, and packet loss. The test setup is illustrated in Figure 7.3.

In the test area, there are five Wi-Fi base stations, whereas 5G technology requires only one base station with one 5G radio cell and five antennas. The Wi-Fi radio is configured for 3 and 7 channels with MIMO configuration, in n78 channel within the 3.5 GHz band for 5G. In underground mine settings, where UL traffic is more important than downlink (DL), the number of UL channels is higher than for traditional DL connections. Nokia is the sole technology provider for 5G radio and SA core networks. In the initial test setup, the control of the loader vehicle used a Wi-Fi (Cisco access point) network, and the test cases were carried out on the 5G connectivity, as there was no other traffic on the 5G network, so there was no congestion in the network. For each throughput test, a background ping test was also carried out. In the second phase of testing, the loader was controlled through the 5G access point, and the same tests were carried out using the Wi-Fi network.

An Iperf3 Server was installed at one end, and an Iperf3 Linux-based client was linked to the 5G and Wi-Fi access points, respectively, to carry out latency and throughput tests. Iperf3 was used to identify the bottlenecks or problems that might occur during a transfer. Iperf3 is a free open-source tool that is widely used for measurements of the maximum achievable throughput between point-to-point connections, and it can be used with transport control protocol (TCP) and user datagram protocol (UDP) protocols. TCP and UDP throughput tests were carried out for each for the same, predefined time. Moreover, ping tests were run to measure packet losses, latency, and jitter. All TCP tests had a default 1400 bytes maximum segment size (MSS).

In a mine environment where real-time video streams are sent from machines and vehicles working underground to a control center on the surface, the UL connection is more relevant than the DL. UL throughput testing for 5G and Wi-Fi was run for TCP and UDP separately, as practical use cases can be configured by the customers using these two alternative protocols in the mines. TCP is connection-oriented, meaning once a connection has been established, data can be transmitted in two directions. It has built-in systems to check for errors and guarantee that data will be delivered in the order it was sent. UDP is the simplest protocol and provides a best-effort datagram service where applications provide their own reliability and flow control. Unlike TCP, UDP is faster but less reliable, which makes it ideal for use cases where speed is more important than data integrity. In the mining industry, it is more common to use UDP than TCP protocols.

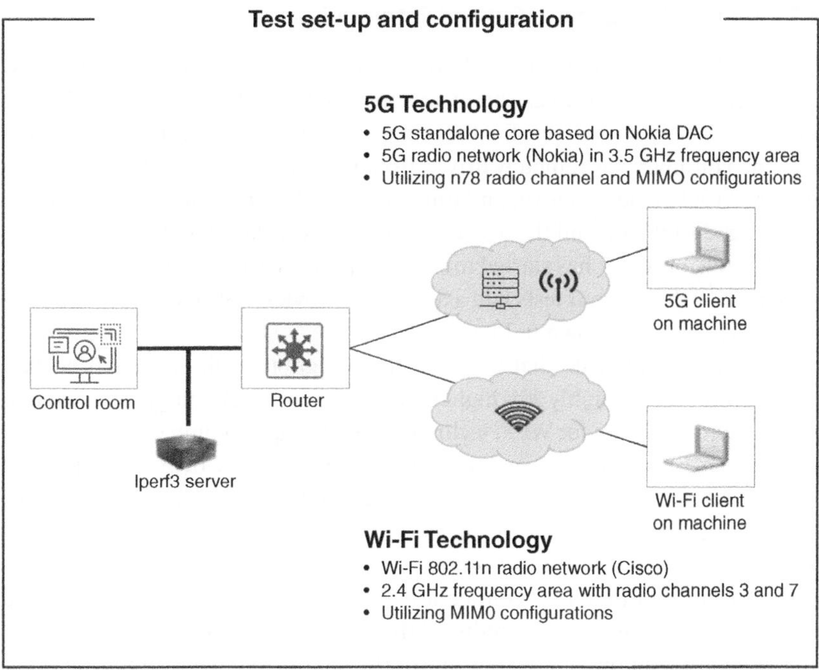
Test set-up and configuration
5G Technology
• 5G standalone core based on Nokia DAC
• 5G radio network (Nokia) in 3.5 GHz frequency area
• Utilizing n78 radio channel and MIMO configurations
5G client
on machine
Control room
Router
Iperf3 server
Wi-Fi client
on machine
Wi-Fi Technology
• Wi-Fi 802.11n radio network (Cisco)
• 2.4 GHz frequency area with radio channels 3 and 7
• Utilizing MIMO configurations

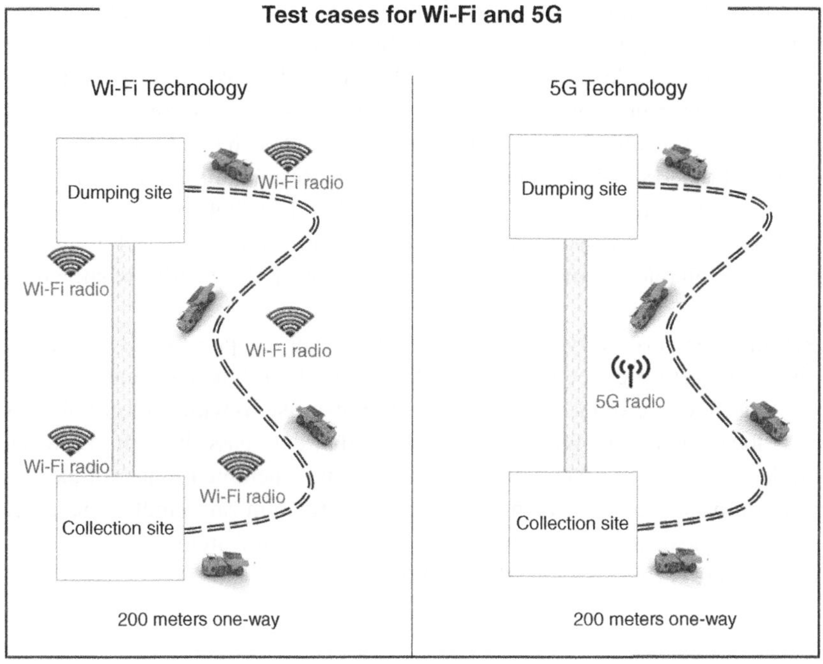
Test cases for Wi-Fi and 5G
Wi-Fi Technology
Dumping site
Wi-Fi radio
Wi-Fi radio
Wi-Fi radio
Wi-Fi radio
Wi-Fi radio
Collection site
200 meters one-way
5G Technology
Dumping site
5G radio
Collection site
200 meters one-way

Figure 7.3 Test setup [5].

7.3.3 Uplink Throughput (TCP) and Retransmission

UL throughput testing using TCP protocol was first carried out with 5G technology. A Nokia 5G access point was installed onto the loader vehicle. The underground testing was run for the same, predefined time during which the unmanned dumper was remotely driven from the control center. The truck had a predefined route to load, move, and unload. After the first test round, the setup was changed to Wi-Fi, and the 5G access point was replaced with a Wi-Fi client. Now, the control center has started to steer the process using Wi-Fi radio technology. Both TCP throughput tests had a default 1400 bytes MSS at each end. A summary of the results is shown in Figure 7.4.

The test results indicate that the average throughput with 5G using the TCP transport protocol is roughly 20% higher than with Wi-Fi technology. The throughput's standard deviation for Wi-Fi technology is 49% higher compared to 5G, demonstrating that industrial 5G provides mining use cases with more stable and predictive bit rates. This enables the use of TCP protocol that offers more reliable connections compared to the existing use of UDP in the mines.

TCP detects packet loss and executes retransmissions to ensure reliable messaging. Packet loss in a TCP connection is also used to avoid congestion and thus produces an intentionally reduced throughput for the connection. In TCP, the congestion window (CWND) is one of the factors that determines the number of bytes that can be sent out at any time. The CWND is maintained by the sender and is a means of stopping a link between the sender and the receiver from becoming overloaded with too much traffic. The CWND is calculated by estimating how much congestion there is on the link. When a connection is set up, the CWND, a value maintained independently at each host, is set to a small multiple of the MSS allowed on that connection. Retransmission is the resending of packets that have been either damaged or lost. Retransmission is one of the basic mechanisms used by protocols operating over IP/TCP networks to provide reliable communication. A summary of the results is shown in Figure 7.5.

Retransmission is a very simple concept: whenever one party sends something to the other party, they retain a copy of the data sent until the recipient acknowledges that they received it. The number of retransmissions with 5G technology is clearly lower than with Wi-Fi. This indicates that data pass through and reach their destination better. Network congestion occurs when the number of packets sent from the source exceeds the number the destination can handle. The larger the CWND, the better the transmission success rate. Based on the test results, the CWND for 5G is almost double that for Wi-Fi.

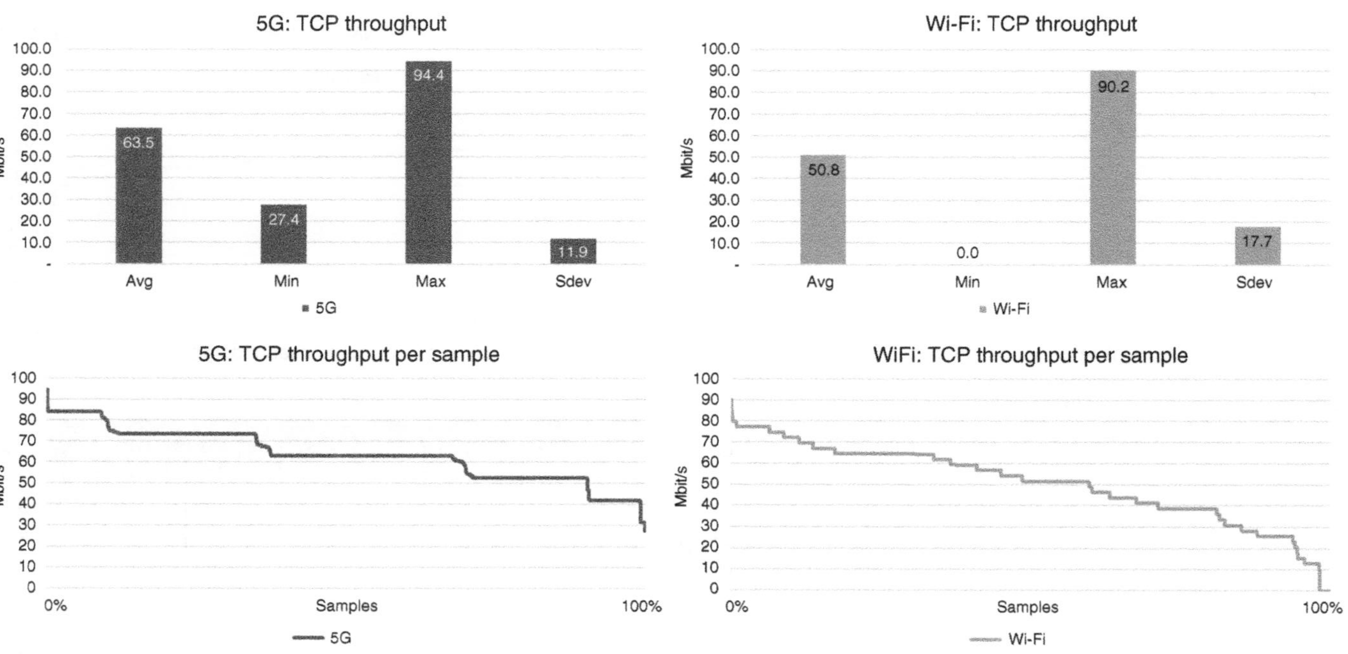

Figure 7.4 Uplink throughput (TCP) for 5G and Wi-Fi [5].

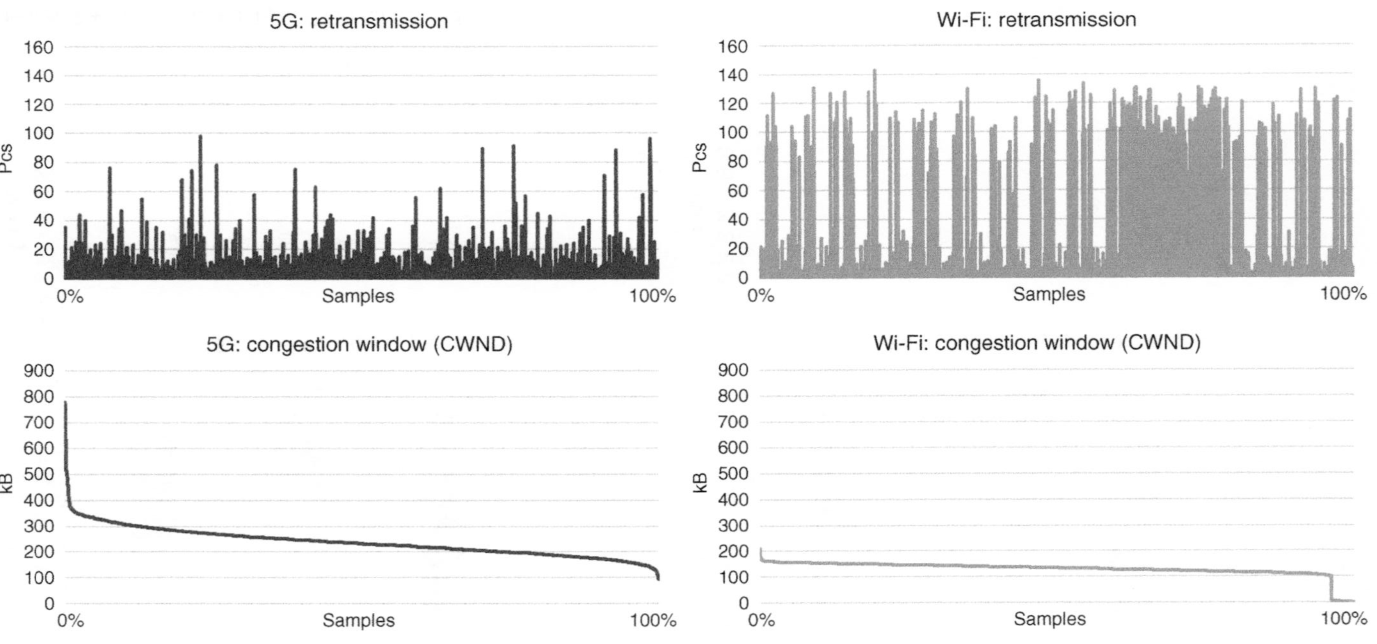

Figure 7.5 TCP retransmission and CWND [5].

7.3.4 Uplink Throughput (UDP) and Packet Loss

In these test results, the performance of 5G and Wi-Fi is compared based on UDP throughput. The upper value of UDP is limited to a certain throughput level in the Iperf3 configuration while performing the tests. In our test case, the UDP throughput was limited to 50 Mbps. In the default configuration, the UDP throughput was limited to 1 Mbps, which was too low. In the other configuration option, where it was set to be 100 Mbps, this clearly produced too many packet losses. Therefore, 50 Mbps throughput was set on the client side to obtain more realistic results. A summary of the results is shown in Figure 7.6.

The test results indicate that the average throughput with 5G using UDP transport protocol is almost equal between 5G and Wi-Fi technologies. There is only a 2% difference in favor of 5G. On the other hand, the throughput's standard deviation for 5G technology is more than half that of Wi-Fi. Also, the UDP test results demonstrate that industrial 5G provides mining use cases with more stable and predictive bit rates.

In UDP, packet losses occur when one or more packets of data traveling across the network fail to reach their destination. It is either caused by data transmission errors in the wireless network or network congestion. A packet loss ratio is measured as a percentage of packets lost with respect to packets sent. A summary of the results is shown in Figure 7.7.

There is a big difference in packet losses between 5G and Wi-Fi, as shown in Figure 7.7. The packet loss ratio for Wi-Fi is more than 20 times higher than that of 5G technology. This is a significant difference and shows how the reliability of industrial 5G dominates the data quality consistency results.

7.3.5 Latency and Jitter

Network latency is an important factor in mission-critical communications. Generally, it is the time it takes for a data packet to go from the sending endpoint to the receiving endpoint. For the case study, we used RTT to compare the latency between 5G and Wi-Fi at the application layer as, for mining applications, it is the total time (data packet from the sending endpoint to the receiving endpoint and back) that matters. The measurements are presented in Figure 7.8.

The test results show that the RTT for 5G seems to be higher compared to Wi-Fi, i.e. latency is lower for Wi-Fi. This can be explained by the fact that the use of Wi-Fi has been optimized over a long time in the test mine, whereas the 5G deployment was recent. The 5G parameter optimization is still to be carried out in the test mine settings. Furthermore, there is a difference in the way RTT is measured for 5G and Wi-Fi. The measurement route for 5G includes the 5G SA core from the base station. On the other hand, the RTT standard deviation

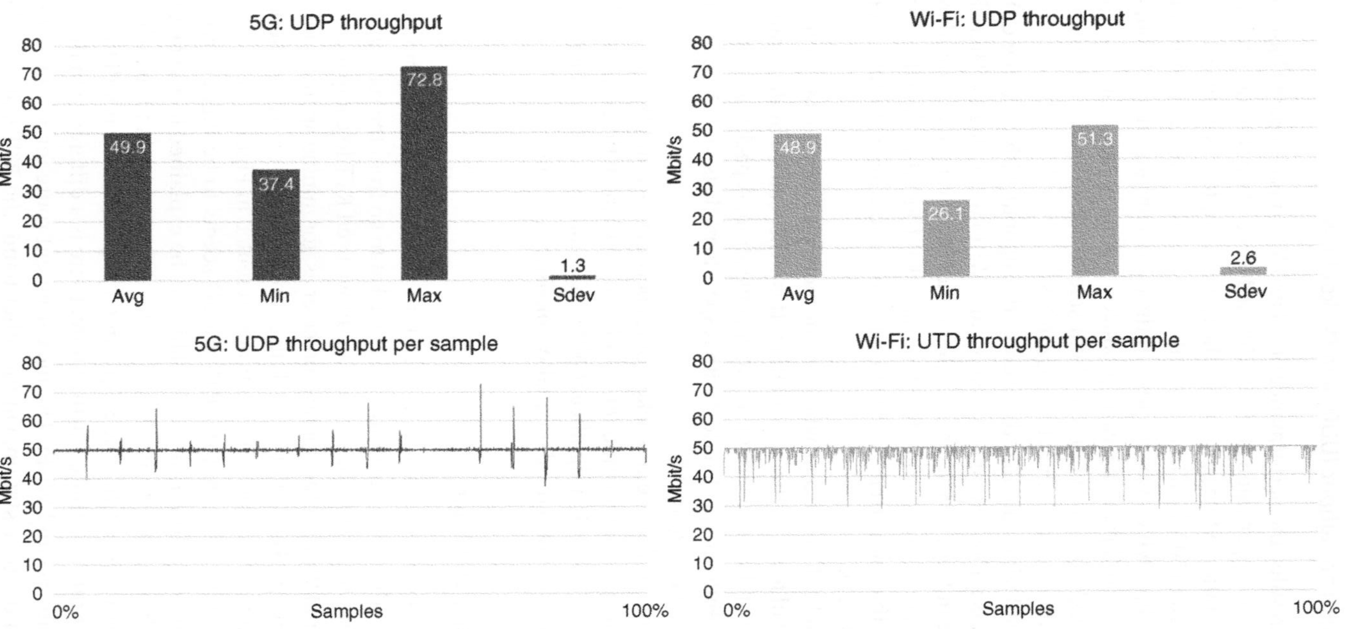

Figure 7.6 Uplink throughput (UDP) between 5G and Wi-Fi [5].

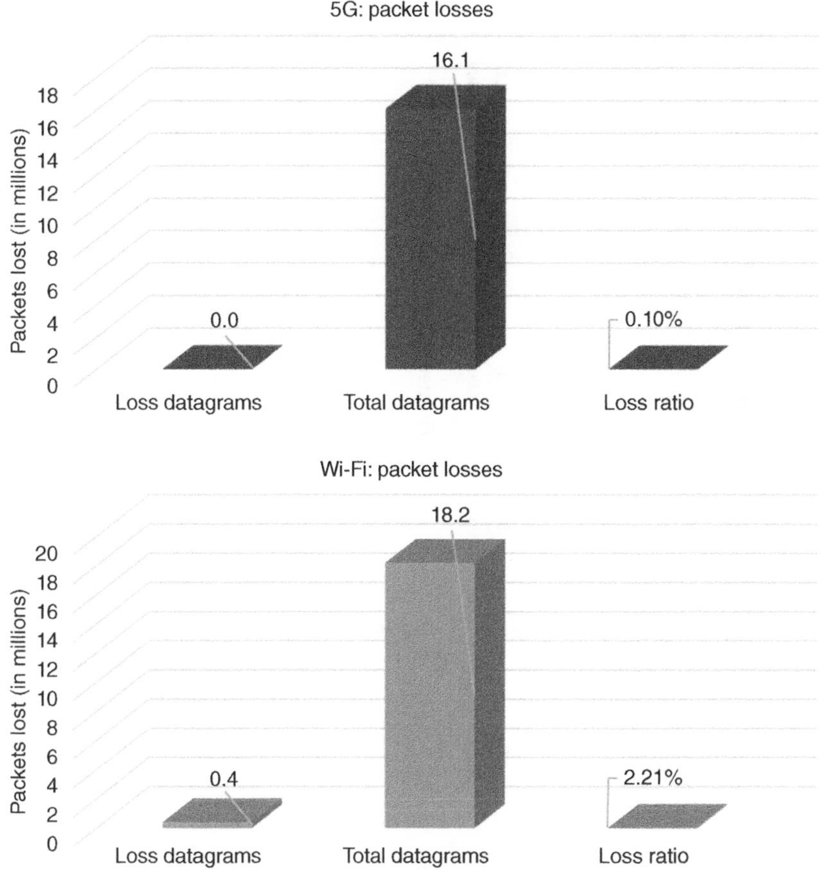

Figure 7.7 Packet losses for 5G and Wi-Fi [5].

for 5G almost equals that of Wi-Fi. This supports the general expectation that, with 5G technology, the latency will be more predictable and stable. Delays that are too long can stop the machine in production due to safety reasons. Feedback from mine operations workers confirms that 5G has not caused any production breaks, whereas they have taken place with earlier technologies. A production break is always a major challenge and can be time-consuming to fix in real mines.

Jitter in IP networks is the variation in the latency of a packet flow between two systems when some packets take longer to travel from one system to the other. Jitter results from network congestion, timing drift, and route changes. A summary of the results is shown in Figure 7.9.

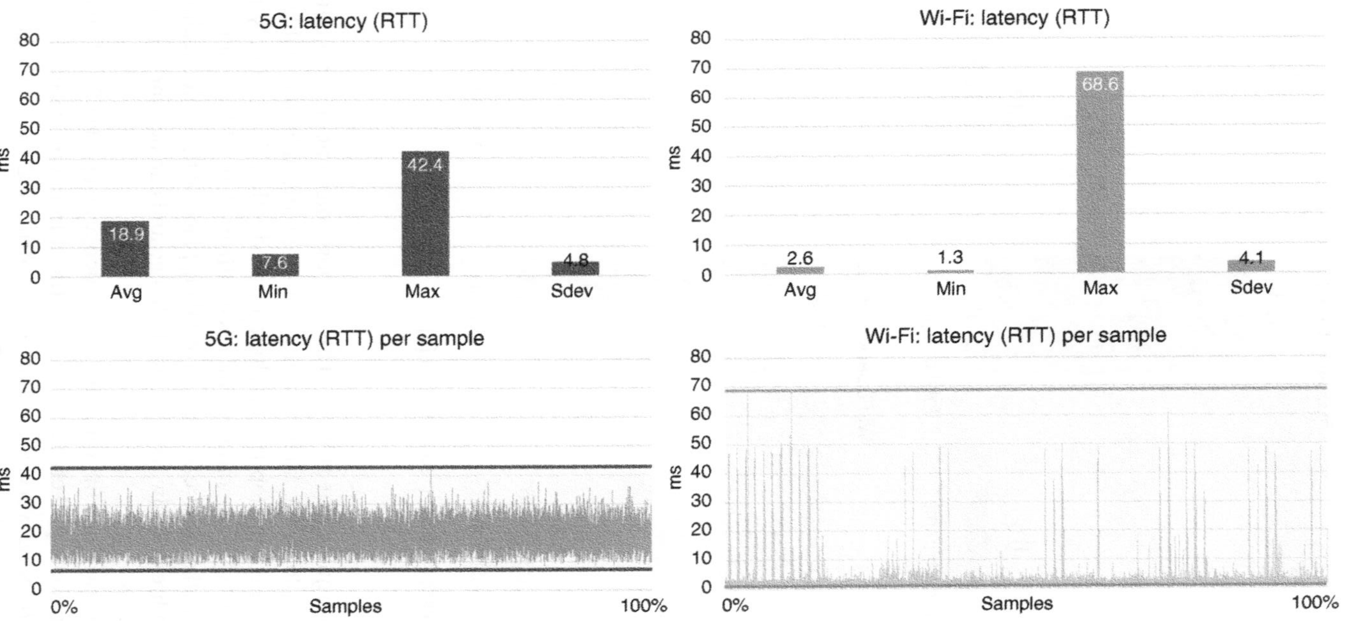

Figure 7.8 Latency (RTT) between 5G and Wi-Fi.

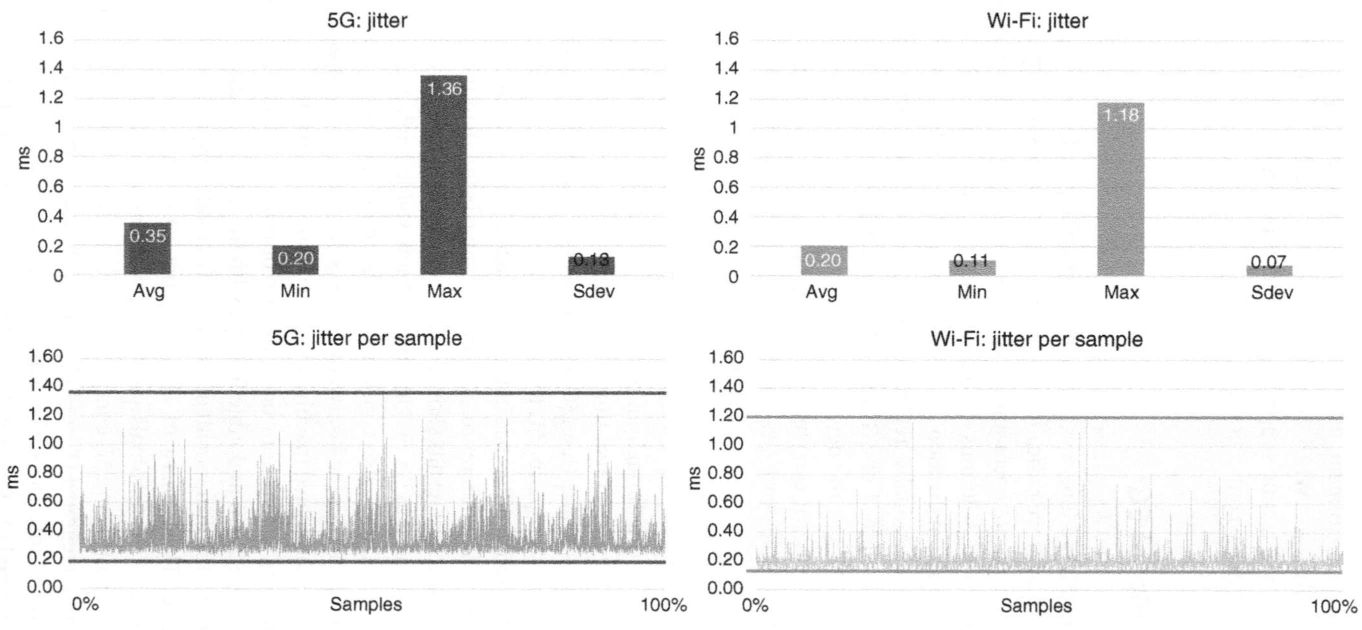

Figure 7.9 Jitter between 5G and Wi-Fi [5].

Based on the measurement, jitter is very low for both technologies. It is 0.35 ms for 5G and 0.20 ms for Wi-Fi. It seems that here, the 5G technology has less deviation than Wi-Fi, thus supporting a general expectation that 5G has more predictable and stable connections.

7.4 Benefits of 5G

To summarize, 5G is a long-awaited mobile technology in the mining industry, as it is expected to improve radically the wireless opportunities underground and on the surface to automate operations. It is, however, not expected to replace all the existing connectivity solutions soon but rather to complement them and allow mines to scale up digitalization at their own pace. The implementation will take time, and it will gradually replace existing technologies for years to come. Nevertheless, the spread of 5G networks triggers a revolution in the industry when all capabilities of the new technology are globally available. 5G opens up completely new ways of carrying out mining operations and improves safety in this kind of hazardous work environment. It is taken for granted that, worldwide, people's safety is a priority for all mines.

Traditionally, cellular networks are designed and built to optimize the DL capacity of the radio channel for users. In the mining environment, when operating underground, it is the throughput of the UL that matters, as machines, vehicles, and IoT applications send real-time information from underground to an operations center on the surface. Here, 5G provides a significant improvement compared to 4G (LTE). When comparing 5G to a well-customized Wi-Fi network, in practice, there is not that much difference in the UL capacity. However, for connection reliability, there is a major difference: 5G is clearly much better than Wi-Fi technology. As the performance test results show, 5G UL throughput and latency deviations are much lower compared to Wi-Fi. This implies more stable mining operations without disturbances. This is also confirmed by the mine operations teams running numerous use cases over the last twelve months: high UL throughput can be fully utilized, and the automated machines work without interruptions.

The new mining applications boost the use of real-time data connections and video streaming. 5G can undoubtedly cope with the increased use of mobile data on the network. However, not all of the new 5G-enabled capabilities and features are available yet. The 5G network alone does not boost digital transformation, but the development of 5G terminals and devices will play a decisive role in the future. Better utilization of secure 5G computing platform capabilities will create opportunities to build integrated OT and IT solutions based on mobile edge computing with standard API interfaces. The network and computing capacity will form an increasingly unified platform for the next stages of mining digitalization.

7.5 Future Opportunities

In the long term, the lag-free nature of 5G will become even more essential for autonomous operations and transport in the mining industry. New real-time applications with future devices and terminals utilizing tactile 5G and the internet will create new virtual opportunities to increase the safety and productivity of mining operations.

Reliable connectivity is key to achieving the business benefits of the big data revolution [6]. Data already drive many industries today, but they will drive all industries in the future. However, collecting data is one thing; it is what you do with it that counts. If our customers are to succeed, they need intelligent equipment that is able to interface with their systems and add true business intelligence [6]. Whatever technology our customers demand – artificial intelligence, image or voice recognition, augmented or virtual reality – it will be underpinned by fast, low-latency connectivity.

References

1 Underground Communications Infrastructure Sub-Committee of the Underground Mining Working Group, "Underground Mine Communications Infrastructure Communications Infrastructure Part III: General Guidelines," Ormstown, Canada: Global Mining Guidelines Group (GMG), 2019.

2 VTT, "VTT, Nokia & Sandvik Collaborate in 5G Powered Research Project on Next Generation Underground Mining Technology," 2023. [Online]. Available: https://www.vttresearch.com/en/news-and-ideas/vtt-nokia-sandvik-collaborate-5g-powered-research-project-next-generation. [Accessed 30 April 2023].

3 M. Timo, P. Daniel and H. Eetu, "Autonomous mobile machines in mines using 5G enabled operational safety principles," *VTT Technology 412*, Espoo, Finland, 31 March 2023.

4 S. Khawaja, "Digitalisation and its impact on the mining industry," *Verdict Media Limited 2023*, London, 17 January 2022.

5 J. Rajani, 5G and Wi-Fi Performance in Underground Mining: Master's Thesis, Aalto University, 2023.

6 J. Vilenius, "Sandvik Director Research & Technology", "https://www.home.sandvik/en/stories/articles/2019/11/a-connected-future/," 5 November 2019.

8

Forest Industry: Improving Productivity in Bioproduct Mill Operations

CASE STUDY TEAM MEMBERS:

Perttu Laiho: Aalto University
Jani Salonen: Metsä Group
Janne Pekola: Metsä Group
Jukka Mokkila: Metsä Group
Jari Collin: Aalto University

8.1 Introduction

The case company for the forest industry is Metsä Group, an international frontrunner in sustainable bioeconomy with strong roots in the Finnish forests. The company produces fossil-free products from renewable wood grown in sustainably managed northern forests. The company invests in growth, innovating new bio-products, and building a fossil-free future for our planet. What truly makes Metsä Group a unique company structure is that the parent company is a cooperative owned by more than 90,000 Finnish forest owners. The company uses the best renewable raw material in the world – northern wood – responsibly and efficiently. Annual wood procurement is 33.9 million m^3 with 100% traceability.

Metsä Group's vision is to be the preferred partner in developing sustainable businesses. Their goal is the advancement of the bioeconomy and circular economy by efficiently processing northern wood into first-class products. The business is based on renewable raw materials and recyclable products, in which wood from northern sustainably managed forests and our profound expertise provide a competitive advantage. Changes in the business environment offer

5G Innovations for Industry Transformation: Data-Driven Use Cases, First Edition.
Jari Collin, Jarkko Pellikka, and Jyrki T.J. Penttinen.

Published 2024 by John Wiley & Sons, Inc.

the company numerous opportunities. With industrial efficiency, the company converts renewable resources into carbon storage and recyclable products for people's daily lives.

The offering of bioproducts covers basic wood supply, forest services, different wood products, sawn timber, pulp, paperboard, and other bioproducts. In recent years, there has been a growing interest in new environmentally friendly product innovations such as biochemicals, biogas, bioenergy, solid energy waste, ash, and lime. Even textiles and 3D fiber products for packaging are emerging innovations for worldwide use. Over the last decade, the company has placed an emphasis on research and development, as the whole forest industry sector is in the middle of a major transformation. The traditional product has been paper in different formats, but due to digitalization, the global paper market has been gradually declining for years, both in demand and production. On the other hand, main global megatrends support the growth of the company. Population growth, urbanization, climate change, loss of biodiversity, and digitalization are the strongest megatrends of the 2020s.

The company defines innovation as being based on knowledge, cooperation, and a new way of thinking. Industry collaboration is important to building winning industrial ecosystems. Metsä Group advances bioeconomies and circular economies by innovating and cooperating beyond traditional business borders. The aim is to find new use cases for our raw material, together with existing and new partners and customers. Together, we can make the world run on renewables. Metsä Group invests in growth by developing technologies and building modern facilities. For example, the Äänekoski bioproduct mill is the world's largest wood-processing plant in the Northern Hemisphere, producing pulp and a broad range of other bioproducts and leading an industrial ecosystem. A similar mill is being built in Kemi.

8.2 Industry Transformation Challenge

The strongest trend in technology is digitalization, which causes changes in consumer behavior, customer experiences, and industry. Digitalization produces a great deal of data for industries, which can be used in production, maintenance, and customer work. It transforms the whole forest industry. In mill settings, digitalization provides a significant opportunity to improve productivity further by increasing overall equipment effectiveness (OEE). OEE is a standard industry KPI for measuring productivity, consisting of three components: availability, performance, and quality. A key question in applying 5G is how to improve the efficiency of bioproduct mill operations in production, maintenance, and safety.

8.3 Data-Driven Use Cases

As forest industry enterprises have continuously improved their operational performance due to increasing competitive pressure from rivals, new process innovations are attractive to ensure future success in operational performance. Most of the core production is already highly automated and, especially in pulp and board, the current trend is toward integrated production lines that rely on wired connectivity due to severe real-time requirements for connectivity in core parts of mission-critical production. However, increasing numbers of tasks will be implemented over the next few years by leveraging industrial 5G to enable new use cases which, in turn, will improve productivity in production. The key to identifying potential use cases is to ensure their technological readiness for a given task and the availability of productivity benefits after the implementation. In practice, it means having a sufficiently robust 5G network for connectivity and necessary complementary technologies, e.g. automated guided vehicles (AGVs), robots, or AI analytics, to create the use case while ensuring the implementation solves a practical problem in the production with clear improvement potential. In addition to the previous two main points, it is also recommended that there be an assessment of implementation risks and issues to ensure that it is possible to carry out the implementation successfully in practice.

From a technological perspective, 5G provides performance improvements in manufacturing and business operations through faster data connections, lower latencies, and a high number of connected devices, enabling real-time analytics and increased automation. The enabler of these changes is the improved technical performance, which was not available in the 4G network, making 5G a multipurpose solution for several communication purposes like mobile on-demand video analytics and remote experts. Other wireless alternatives, such as WLAN and WPAN, have typically lower data rates, bandwidth, and capacity but can cover longer ranges with lower deployment costs and fewer skills needed from the workers operating the network. The drawbacks of most of the alternative solutions are their limited data rate and varying reliability, which limit their use cases, especially in Industry 4.0 applications like autonomous vehicles and AR solutions. The upcoming 5G releases 16 and 17 are also significantly improving the capabilities of 5G by introducing more features supporting industrial automation. This development suggests that 5G can outperform other solutions, but it comes with high costs and advanced know-how requirements for the network operators. Despite these drawbacks, 5G has the potential to provide significant improvements for industrial companies by enabling new opportunities for digitalization. This enables performance improvements and ways to organize working, which existing industrial networks or 4G can't provide.

Table 8.1 summarizes three potential use cases for a forest industry [1] production plant based on the case study carried out, after which this section elaborates on each use case to clarify its characteristics and productivity impact. It is important to note that these three use cases are mostly "conceptual" at the time of writing

Table 8.1 Forest industry use cases for productivity improvement.

	AGVs for process automation	**On-demand site monitoring**	**Digital maintenance support**
Description	Automated guided vehicles for production processes to replace human labor in repetitive tasks, especially in logistics	Automated and portable wireless monitoring systems combined with AI analytics analyze the data and suggest actions for employees	Real-time access to remote expert support and advanced digital materials (e.g. AR content or video instructions)
Type of benefit	• Increased process efficiency • Better prevention of injuries and accidents	• Better prevention of injuries and accidents	• Reduced downtime • Better prevention of injuries and accidents
Magnitude of benefit[a]	• Efficiency: +++ • Safety: +	• Safety: +	• Downtime: ++ • Safety: +
Implementation time	Software and connectivity updates for existing hardware take one to three years but new hardware requires greenfield or renovation projects that take three to five years	Camera implementation within one to two years and adding robots to carry them two to five years	Most of the digital materials and remote experts are one to three years old and AR content is three to five years old
Examples of implementation based on the case study	Autonomous log yard crane for unloading trucks and trains for pulp mills or autonomous crane for pulp bale loading to trains	Wireless cameras monitor the use of safety gear or ad hoc locations, and robots monitoring, e.g. gas leakages or equipment performance	Portable video camera and phone to consult system suppliers' maintenance experts and AR content to guide, e.g. electrical installations and repairs

[a] To rate the magnitude of the productivity benefits, use cases were compared to each other based on interview data. Rating "+++" indicates significant potential to improve productivity in relation to other use cases, "++" indicates medium potential, and "+" indicates small potential. The assessment is qualitative.

this book because forest industry production plants have only recently begun to pilot 5G-based use cases in their production. Only one practical example exists in the data of on-demand site monitoring, which uses 5G-based video cameras to support communication and monitoring of the site or equipment on an ad hoc basis. Even though the 5G connectivity worked well with the video camera, there was no opportunity to gather exhaustive case evidence from the already implemented use cases, forcing the focus to be on use case potential in the near future. This aspect is highlighted in Table 8.1 as well, to provide evidence of the estimated availability of scaled implementations.

8.3.1 Automated Guided Vehicles

The use of AGVs in process automation is based on changing individual parts of the production process to replace humans with automation to increase the efficiency of production. For Metsä Group, these changes are targeted at the inbound and outbound logistics, as they benefit the most from increased automation, such as automated log yard cranes and pulp bale loading on trains. This is because other parts of the production process are already automated, and an additional improvement may not be financially viable due to high implementation costs but only a limited increase in process efficiency.

In contrast, inbound and outbound logistics are, to some extent, undertaken by humans. This makes the estimated efficiency increase higher, justifying the investment even though the investment costs remain relatively high as well. Overall, AGVs were estimated to provide significant productivity improvements through increased efficiency in the given production phase because logistics tasks are very repetitive and automation systems usually carry them out more efficiently than humans. Based on the interviews at Metsä Group, the productivity improvement was preliminarily estimated to be between 0.1% and 1% for a production phase using AGVs instead of human-operated machines [1]. Due to the large scale of the forest industry, the absolute financial benefit was seen enough to consider these use cases a viable option, even though precise profitability estimates were not available. However, these estimates are only preliminary and have a very high uncertainty as they have not been validated in a real-life experiment but rather are considered as a potential magnitude of improvement.

Another benefit is improved safety because AGVs reduce the personnel in locations with poor ergonomics or with significant risk of injury, such as warehouses. The link to productivity was considered an indirect one since employees without injuries spend more of their time on their primary tasks instead of being on sick leave, which is usually costly for the employer. Also, better ergonomics were perceived to increase productivity by enabling a better focus on the task itself instead of the challenging working conditions. Improved safety was seen to support Metsä

Group's "Safety comes first" policy and was considered worth further investments despite overall good working conditions and low injury rates at Metsä Group.

Despite the large magnitude of the benefits, AGVs' implementation can be hindered due to two reasons of which the first is the availability of suitable equipment. Most forest industry automation suppliers are laggards in the adoption of digital technologies, which can hinder the adoption of 5G-based automation solutions. This is because process industries' automation suppliers typically prefer high reliability and predictable features of the well-known legacy technologies instead of radically aiming to leverage the newest alternatives, such as 5G.

The second reason is the timing of the project. In practice, this means that each larger implementation project must be done outside of the production phase of the facility, that is, during a renovation phase or a greenfield project. This is because critical production process components are not changed during the production phase in the process industry as the facility is planned to work according to predefined parameters for the entire production phase. If all facilities are in the production phase, AGVs' implementation will be moved into the future, thus delaying the benefits. Hence, automation projects based on 5G should be considered as a long-term option for 5G adoption while having a focus on smaller and simpler use cases to achieve the quick first benefits and build skills related to 5G technology.

8.3.2 On-demand Site Monitoring

On-demand monitoring concerns the introduction of portable and automated wireless monitoring systems that use AI to analyze the data and recommend actions, for example, to process operators or shift managers in a factory. The key is to use wireless systems in locations that cannot be wired or where it is inconvenient to do so. On a practical level, wireless cameras could monitor the use of safety gear in a maintenance location otherwise not covered by wired monitoring, or process operators could monitor ad-hoc locations when equipment showed signs of unexpected behavior. The data from these solutions would be analyzed by AI to minimize human effort in watching, for example, a video stream instead of focusing on only the potential issues the AI suggests. A more advanced option would be using small robots to monitor accidents such as gas leaks to avoid risking employees. Another example is to measure systematically a large number of targets such as the bearings of a conveyor belt typically used by the forest industry to transport wood chips as a part of pulp production. Robots would carry out these kinds of tasks more reliably than humans, who consider such tasks boring, which has been seen to affect the reliability of the measurements and thus the performance insights generated by the AI, such as a risk of fire due to overheating bearings.

On-demand monitoring provides an indirect productivity benefit as it focuses on reducing negative events such as accidents or machine failures through better analytics and more extensive data gathering. In practice, this means enhanced monitoring of, e.g. chemical leakages and the use of safety equipment. The rationale for preventing equipment-related negative events is that such prevention increases productivity by reducing the expensive downtime of the equipment during which it is maintained. For human-related injuries or accidents, productivity improvement comes through reduced sick leave as employees use more of their time carrying out productive tasks instead of generating costs without providing the corresponding benefit to the employer that is the case during sick leave. Even though the benefits were recognized, their magnitude was difficult to estimate as negative events were usually rare and continuous monitoring has only a small impact on productivity, for example, through incremental improvements in certain process parameters when poor quality is detected. However, the benefit was seen to be sufficient on a general level, as preventing only one larger equipment failure or accident is likely to pay back the investment in the monitoring system and prevent the loss of expensive production time or talented employees, even though exact quantification was perceived too unreliable due to the high case specificity of the benefits.

Implementation time depends on the specific use case and the required technologies. Shorter-term implementations, such as portable cameras and related analytics, are possible within one to two years as the necessary technologies are already available and analytics have been found to be reliable in other similar implementations, suggesting that recommendations would be reliable. Adding autonomous robots to carry out monitoring tasks requires two to five years, making it a longer-term option as robots were perceived as unreliable and expensive for use in factories. However, with sufficient technologies, the implementation was not perceived as an issue, as Metsä Group was seen to acquire the monitoring as a service rather than develop extensive knowledge regarding the maintenance and installation of the robots. Relatively fast deployment of a monitoring system was seen as possible for both robots and simpler non-robotic solutions. This is because these systems would not control the mission-critical process and, hence, the deployment is possible without major renovation or greenfield projects that are, in general, rare due to the long production phase of a forest industry factory.

8.3.3 Digital Maintenance Support

Digital maintenance support concerns the introduction of digital tools for various maintenance tasks to increase maintenance efficiency, that is, to decrease the time needed to complete a given maintenance task. For Metsä Group, two typical examples were found. First, maintenance workers could use 5G-based video cameras and mobile phones to seek assistance from, e.g. automation system suppliers' experts remotely when carrying out an unfamiliar task. Through direct video and a high-capacity data connection, an expert can provide advice and instructions without first needing to travel to the factory, which usually takes a long time due to its remote location. Second, maintenance workers could access more advanced instruction materials, such as AR content, to help them with electrical installations. This makes following the instructions more convenient and could provide suggestions of best practices for carrying out a given maintenance task.

The previous examples also demonstrate the benefit of digital maintenance support for decreasing downtime by improving maintenance performance. This is a direct productivity benefit, as lost production time is expensive in the forest industry, and efforts to minimize it directly reduce the cost of maintenance periods and unpredicted equipment failures. The magnitude of this benefit was considered medium, as a few hours shorter downtime is typically worth tens of thousands, according to the interviewees, even though they emphasized that the exact saving depends on the pieces of equipment that broke down and only general-level estimates could be provided. On top of the direct productivity impact, better digital materials were found to improve occupational safety during maintenance projects by providing instructions covering the best practices for doing a task. This, in turn, leads to a reduction in injuries and sick leave, which improves productivity by reducing the idle time of the workers. However, the overall impact is rather small, and the main impact of this use case comes from the potential to reduce equipment downtime.

The implementation timeline is dependent on the specific implementation. Most of the digital materials and the remote expert service can be implemented within 1–3 years, as the required technologies such as video cameras, 5G connectivity, and 5G-based mobile phones are already available and likely to be more popular over the next 12 months. Also, other supporting factors, such as making the necessary service deals with equipment suppliers, are possible to achieve as they were considered as extensions to the existing maintenance agreements. However, AR technology was estimated to take three to five years before it became robust, light, and convenient enough to use in factories. This limits the availability of advanced digital materials such as AR-based wiring instructions for electrical installations. Another important note is that maintenance support systems are relatively simple to implement, suggesting that when the necessary technologies are available, implementation is fast and the benefits are quickly available. This is because maintenance support implementations do not require a change to the critical manufacturing process and is usually changed only during renovation or greenfield projects.

8.4 Benefits of 5G

8.4.1 New Opportunities

While the last section focused on the recognized use cases that could improve productivity, it left unanswered questions regarding the link between 5G networks and productivity. Hence, this section summarizes the productivity benefits available for forest industry enterprises on a general level and discusses the relationship between 5G networks, complementary technologies, and productivity benefits.

The benefits are determined based on the use cases, but they can also be considered as the productivity benefits of 5G at a more general level, even though 5G networks alone were not found to improve productivity. This is because 5G played an important part in enabling the recognized productivity benefits. The case-level benefits can be categorized into (1) prevention of injuries and accidents, (2) reduced equipment downtime, and (3) increased process efficiency. A uniform benefit available through all 5G-based use cases is the prevention of injuries and accidents due to, for example, a reduced need for working in dangerous locations such as tall warehouses or the potential to have better instructions related to daily tasks using digital tools. However, the impact on productivity was rather small as improvements in safety were seen as difficult to convert to significant improvements in productivity as, for example, sick leave is already rather rare due to injuries.

In contrast, reduced equipment downtime through maintenance support and, especially, increased process efficiency through AGV use can improve productivity significantly. This is likely because both can directly influence critical parts of the production. A small time-saving of a few hours in a maintenance task or a small relative improvement in the production volume results in a large absolute increase in the produced quantity due to the large scale of forest industry production plants. As Table 8.1 demonstrates, the largest productivity benefits depend on the specific use case, suggesting that all 5G-based use cases might not significantly improve productivity even though, in general, 5G use seems to have clear productivity benefits for forest industry production plants.

There have been a few trials to test how 5G technology can increase productivity and safety in a mill area. 5G use cases for digital maintenance support, AGVs for process automation, and on-demand site monitoring are briefly described in the next sections.

8.4.2 Using the Frans 5G Robot to Explore Digital Maintenance Opportunities

A promising example is the robot dog Frans, which participated in an experiment carried out by the company at its traditional pulp mill, conceptualizing future service robotics. The experiment was part of the MURO robotics project facilitated by the VTT Technical Research Centre of Finland, whose aim is to find out what kind of added value a service robot can bring to factory work [2]. Metsä Group wanted to explore the possibilities of utilizing a service robot in a mill environment. The experiment examined what kind of maintenance and service tasks the Frans robot can carry out with the current technology and, for example, free up the time of the people working in the factory for other kinds of tasks.

Frans' operations in a genuine factory environment produced valuable information on the direction in which robotics should be taken in industrial operations and how robots can best work together with humans. The ability of robots to collect and process data opens completely new doors for securing and streamlining industrial operations.

The robot dog used the 5G network to collect and produce information about its observations in real-time, which is essential for maintaining safe industrial operations. Frans used its 3D camera to carry out scans that support, for example, the planning of modifications. In addition to this, the robot was sent to measure the temperatures of various components, such as individual valves, with its thermal imager. In the future, robots may also be used to minimize dangerous situations: observations of gas leaks or overheating, for example, need to be done both quickly and safely. Frans is shown in Figure 8.1.

Figure 8.1 Service robot Frans at a pulp mill. *Source:* Reprinted with permission from Metsä Group.

The Frans robot dog represents the latest robotic 5G-enabled infrastructure, which means that robots can already be used for certain automated functions in addition to working with humans, for example, in industry maintenance tasks. Maintaining the operations of a pulp mill involves a wide range of tasks, some of which can be challenging, requiring extreme precision or repetitive work for a person. Navigating hot and loud spaces alone is easier for a robot than for a human. One of the goals of the experiment was to divert repetitive and risky tasks to Frans. These included, for example, scans and measurements for detecting problems. At the same time, it was seen whether the robot could move across factory structures, such as tricky staircases, without outside help.

Frans the robot dog utilized, among other things, 3D technology, i.e. lidar scanner, thermal imager, 5G network, and edge computing, when moving around the premises independently and collecting information. Frans traveled in the factory environment and carried out various measurement and scanning tasks important for the factory's operations. The results of the experiment provided an understanding of how 5G service robots can be utilized in factory operations and how they can contribute to, for example, improving occupational safety and efficiency. The results of the experiment were promising and showed what kind of essential functions a 5G robot could perform in a factory environment with the help of humans, perhaps even better than a human. On the other hand, certain functions are not yet supported by the current functions of the robots, so their testing and development is still ongoing. However, the experiment provides new insight into how to improve the mill's OEE even further.

8.4.3 Transferrable 5G Video Cameras for Site Monitoring

Another interesting 5G use case to improve OEE is a portable and easily transferrable wireless video camera solution with real-time video analytics capabilities running on the mobile edge. The video camera can monitor a specific process, business environment, material flow, or moving people while meeting privacy obligations. There are some processes where temporary monitoring for a specific period is needed. These include vehicle collision detection, PPE compliance, area surveillance with behavior models, hazardous materials monitoring, material leak detection, fire hazard prevention, and security monitoring.

The 5G use case utilizes an easily deployed virtual private 5G network with a regional mobile edge computing center. Figure 8.2 shows the technical setup for the use case.

A 5G onsite mobile application is a digital tool for industries. It enables ad hoc video meeting services optimized for industrial field work. It is web-based (runs on smartphones, tablets, and PCs) with no installation – it works by just opening an invitation (a web link). It supports the use of multiple cameras and easy swaps

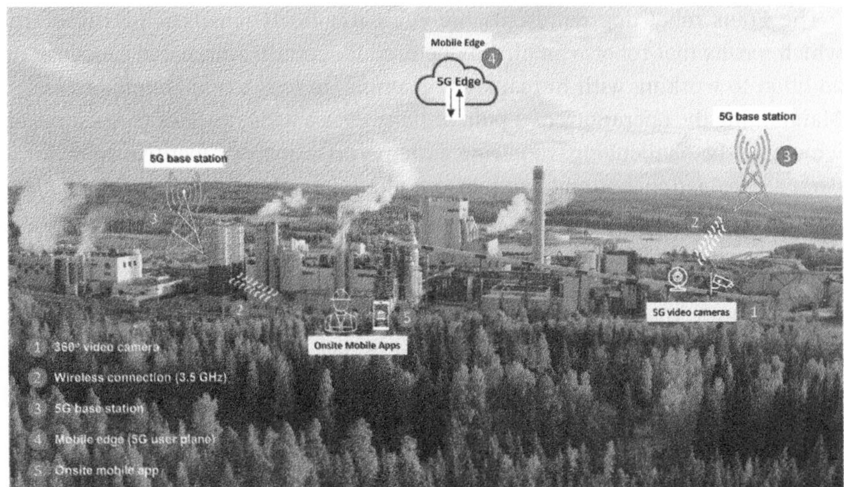

Figure 8.2 Transferrable 5G video cameras for site monitoring. *Source:* Reprinted with permission from Metsä Group.

between cameras, as well as optimizing resolutions and frame rates. Sharing of documents, desktops, and pictures in joint collaboration meetings is possible.

By using the transferrable wireless 5G video camera solution with video analytics, it is possible to provide monitoring of an AI-supported video stream to a mill's control center, and in real-time integrate it into the rest of the processes controlling production. It is a viable, temporary solution to allow a focus on certain non-optimal parts of the process and provide the operators with fact-based insight on how to fix the issues.

8.4.4 Enhanced Automation for AGV-enabled Processes

Currently, AGVs are being used in different parts of the pulp production process in a mill area. In the first part of the process taking place inside a mill area, there are unmanned, remotely controlled high-lift wheel loaders that unload wood from the trucks. As big trees can randomly fall from the lift, it is classified as a dangerous area for people to move in. Therefore, semi-automated AGV solutions are primarily used in these unloading locations. Fallen trees are not always noticed and can cause an unwanted stop to the process, damaging vehicles or rails from the unloading area to the debarking location (Figure 8.3).

Machine vision is a 5G-enabled use case that recognizes hazardous situations and safety deficiencies by using real-time video analytics integrated into the unloading process through the local control center. Machine vision as a

Figure 8.3 "5G machine vision" trial in the wood unloading process. *Source:* Reprinted with permission from Metsä Group.

real-time application is based on 360° cameras, which constantly identify objects, moving people, and any anomalies in the danger area. Using a 5G connection, this information is sent to a private cloud, where an inventory map of the field is updated. Machine vision solutions can identify whether objects have been damaged during handling and then store that information. Sensor information about the storage unit (temperature, humidity, etc.) can be utilized. The location of all objects is known at all times, and further transportation phases will receive exact locations for loading. There is no need to search for missing objects, which reduces disturbances, search operations, and waiting time for trucks.

8.5 Barriers Hindering Adoption

As the recognized use cases were mostly found to have potential for the near future, it is necessary to consider the likely barriers that might hinder their implementation. This helps the forest industry enterprises to identify better likely key bottlenecks in the implementation and prepare accordingly to remove such barriers. A summary of the barriers is presented in Table 8.2.

Table 8.2 Barriers hindering the adoption of 5G-based use cases [1].

Barrier category	Common barriers	Additional use of case-specific barriers
Technological readiness	• Limited coverage and reliability of 5G networks in industrial sites	• Lack of advanced complementary technologies (e.g. AR, monitoring robots)
Personnel's willingness to adopt	• Change resistance toward changing practices and new technologies	• Strong involvement of unions if personnel find the change to risk their jobs or do not accept increased monitoring at the site
Organizational knowledge	• Lack of skills to efficiently use the 5G-based use cases	• Rarity of competent cross-functional teams to develop AI-analytics for industrial processes
Access to key partners	*No common barriers*	• OT system suppliers are reluctant to adopt digital technologies • Fragmented solution offering in monitoring systems • Difficulty in synchronizing maintenance offerings of various suppliers
Systems integration and implementation	• Poor IT/OT integration of forest industry companies • Necessity of large ICT upgrades to large-scale 5G-based use cases	• Difficulty in reconciling physical processes to function smoothly with a new automated process phase
Management support	• Existing organization of work does not support digitalized working • Low number of high-quality training courses	• Necessity to establish new units to manage digital maintenance materials • Current data sharing policy limits maintenance support use cases
Financial viability	• High cost of robust large-scale 5G implementations	• Difficulty in demonstrating the financial benefits of on-demand monitoring

At a high level, there are two types of barriers: common and case-specific. Common barriers were found to exist regardless of the specific 5G use case being implemented. For example, limited coverage and reliability of the 5G network in industrial settings are fundamental problems emerging due to the wireless connectivity of 5G being sensitive to physical objects that are very common in industrial sites. Another important barrier is the cost of large-scale 5G implementations, as they typically involve building expensive private 5G equipment for the target site. As both are external to the adopting company, there is very limited opportunity to affect them. Possible approaches could be co-development with key actors in the 5G ecosystem to foster the development or simply wait for more affordable and robust alternatives if existing solutions are not sufficient.

The rest of the common barriers were found to be internal to the adopting company, implying that the company can affect them to aid the adoption of a use case. For example, at present, poorly integrated Information Technology (IT) and Operational Technology (OT) systems prevent efficient and safe integration of new digital technologies. This is because legacy OT systems are not designed to be interconnected due to their poor cyber security. Furthermore, IT and OT systems are always customized in forest industry plants, so nonstandard programming interfaces can complicate the integration. A closely related barrier is the need for significantly large changes in the Information and Communication Technology (ICT) system to enable large-scale 5G implementations, which are necessary to prevent security breaches and enable smooth functioning of the adopted use case. However, although these changes are internal, they are typically complex and expensive to implement. Hence, they are mostly carried out during renovation or greenfield projects to minimize the threat of interrupting the production process due to an unpredicted cascading effect through the system. As renovation and greenfield projects occur rarely for each site, the adoption may be delayed for several years until such a project takes place according to the planned life cycle of the production site.

Resistance toward changing working practices and technologies comes from individual employees and is a typical phenomenon seen in innovation adoption. Overall, change resistance was not considered a major issue as the personnel of Metsä Group were described as being flexible in the sense that previous changes had already introduced new practices and technologies. This is mostly due to the employees' ability to participate in the adoption effort and their general understanding of the need to adopt digital technologies to improve performance in production plants. There are also two related barriers: an organization-level lack of skills to use 5G-based systems and a low number of high-quality training courses. The former is a logical necessity when adopting new technologies as, without proper skills, they cannot be used efficiently. This is especially relevant for those employees who are not technologically savvy but would have to use the new technology as part of their daily work after the adoption. The latter is the current lack of high-quality training courses for new digital technologies, which was considered to hinder adoption and efficient use in Metsä Group. As this barrier was recognized for systems adopted in the past as well, it is likely that completely new technology will have similar issues if no corrective actions are taken.

Finally, the existing organization of work was deemed unsuitable to support efficient use of 5G-based technologies. Thus, it seems necessary that, for the adoption to be successful in providing productivity benefits, changes to the organization are necessary to align the new 5G-based use cases with efficient ways to use them. This is because, for example, the goal of Metsä Group is to extract productivity benefits from the adoption rather than focus on technological novelty without benefits. An important note is that, as the internal barriers are specific to a company, it may be that different barriers have differing criticality if another company is studied. In contrast, the external barriers are either technology- or ecosystem-specific, implying that they could be similar for at least other similar companies such as other Finnish forest industry enterprises.

In addition to common barriers of 5G adoption, there seems to be a set of various but use case-specific barriers that occur depending on the analyzed use case. This seems a logical outcome, as 5G's productivity benefits require both the 5G network and a complementary technology leveraging of 5G to complete a task. Hence, use case-specific barriers are fundamentally the barriers to the adoption of the complementary technology rather than a direct hindrance to adopting the 5G network. However, due to the need to use both 5G and its complementary technology to achieve productivity benefits, it is necessary to recognize the existence of these barriers even though no common patterns were present to support generalization to other use cases. This suggests that companies adopting 5G-based use cases to improve productivity should consider the specific characteristics of each use case from a barrier perspective instead of relying on only the recognized common barriers. Also, it should be noted that the case-specific barriers can be either external or internal. The internal barriers can be controlled by the company itself, whereas the external barriers depend on other actors in the ecosystem or the selected technology, making them more difficult to remove.

Finally, despite several recognized barriers hindering the adoption, critical barriers did not exist, suggesting that 5G, in general, may soon be adopted to provide a basis for productivity improvements through new use cases. However, use case-specific barriers can still be critical and difficult to remove to the extent that they block the adoption of the use case entirely. For example, the main concern for using AGVs is likely the reliability of 5G, which was perceived as a critical barrier as it can compromise the smooth functioning of the entire use case. Also, as it is an external technology barrier, the adopter has very limited opportunities to affect it. In contrast, currently unsuitable data-sharing policies are internal and were considered relatively straightforward to change by the company's management.

8.6 Managerial Implications

As demonstrated earlier, 5G enables new opportunities to transform forest industry production processes digitally and access new productivity improvements through novel use cases. There are four main implications for forest industry managers.

First, forest industry enterprises should focus on video-based on-demand monitoring and maintenance support as the first phase of 5G adoption, as these provide some benefits while being readily accessible due to easily solvable barriers and available technologies. When sufficiently mature complements emerge for AR and monitoring robots, forest industry companies can more easily proceed to adopt them as common organization-related barriers have already been solved.

Second, AGVs should be considered as a future option for inbound logistics for logs and outbound logistics for pulp bales to increase the level of automation and, thus, efficiency in the production process.

Third, to aid the adoption of 5G-based use cases, companies should focus on removing the common barriers, as these hinder the adoption of all 5G-based use cases. Then, a more case-specific focus can be taken based on the use cases a company aims to adopt.

Fourth, technology and viability-specific barriers are difficult to remove by the actions of an individual company, as both are fundamentally about the selected technology, which is beyond the direct influence of forest industry enterprises.

Overall, it is clear that 5G enables important productivity benefits for the adopting forest industry enterprise but comes at the cost of several barriers that must be removed to enable the adoption. Even though this study proposes direct avenues for implementing 5G in forest industry, the technology is still new, and entirely new opportunities are likely to emerge in the future.

References

1 P. Laiho, 5G-enabled digital transformation in the Finnish forest industry, Espoo: Master's Thesis, Aalto University, 2023.

2 Telia, "Robottikoira auttaa vaativissa tehtävissä: "Ihminen kiittää ketterää työkaveriaan"," *Helsingin Sanomat*, vol. 5, p. 9, 2022.

9

Elevator Industry: Optimizing Logistics on Construction Sites with Smart Elevators

CASE STUDY TEAM MEMBERS:

Janne Öfversten: Kone
Ella Koivula: Aalto University
Mika Kemppainen: Kone
Tommi Loukas: Kone
Jari Collin: Aalto University

9.1 Introduction

The case company for this chapter is KONE, a global leader in the elevator and escalator industry with a mission to improve the flow of urban life. The company provides elevators, escalators, and automatic doors, as well as solutions for maintenance and modernization, to add value to buildings throughout their life cycle. Using the more effective KONE People Flow®, the ambition is to make people's journeys safe, convenient, and reliable in taller, smarter buildings. In 2022, KONE had annual sales of EUR 10.9 billion and, at the end of that year, over 60,000 employees. Revenue includes products and services in all phases of the building's life cycle: new equipment (54%), maintenance (33%), and modernization (14%). KONE operates in more than 60 countries around the world, serving over 550,000 customers. Headquartered in Helsinki, Finland, it has eight global R&D centers and ten manufacturing units in seven countries, as well as a worldwide network of agents and authorized distributors.

KONE offers innovative and eco-efficient passenger and goods elevators for all types of buildings, from low and mid-rise structures to the world's tallest

5G Innovations for Industry Transformation: Data-Driven Use Cases, First Edition.
Jari Collin, Jarkko Pellikka, and Jyrki T.J. Penttinen.

Published 2024 by John Wiley & Sons, Inc.

skyscrapers. The company provides solutions that ease the flow of people and goods in new buildings, as well as those where the elevators need modernizing and existing buildings without elevators. Creating good people flow takes a combination of solid data and a deep understanding of how people move around a building to optimize the people-flow design. Nowadays, People Flow® solutions utilize more and more data analytics and machine learning to create buildings with seamless people flow and insights into what works and what does not – and why. Decisions based on accurate data mean a better user experience for everyone.

The growth of the elevator and escalator industry is driven by three key megatrends: urbanization, sustainability, and technology. As the urbanization megatrend shows, the world's cities are constantly growing. They attract billions of people and, by 2050, more than two in every three people on the planet will live in urban areas. Estimates tell us that around 200,000 people move into cities across the globe each day. Urbanization continues to boost the construction of tall buildings for housing, offices, and public services. Elevators are needed throughout the life cycle of these buildings. KONE's goal is to provide ease, effectiveness, and experiences to users and customers over the full life cycle of buildings. The best experience can be created by working together with customers and partners at every step of the process, from early engagement to upgrading equipment. Already today, KONE is a part of the building management from the time of purchase of the land to construction and through the lifetime of the building.

As buildings become taller, their construction and implementation also require elevators during the construction phase. Already in the early stage of construction, elevators are needed to lift construction material, the workforce, heavy tools, and machines up and down between the floors. Therefore, an elevator is, typically, the first "smart" component that is at the center of logistics processes on construction sites. A huge amount of operational data to optimize material, people, and information flows inside and between the buildings can be collected from elevators. KONE strives to optimize elevator usage by creating a common, real-time data platform based on application programming interfaces (APIs). Using standard APIs creates many opportunities: data exchange with other construction stakeholders (subcontractors, technology providers, logistics companies, etc.), fast application development, management of the digital ecosystem, and many others. The company has developed a solution concept called KONE SiteFlow to discover new business opportunities and expand its offering in this area.

Smart elevators can significantly improve the productivity and safety of construction projects. A good example is KONE JumpLift, which is an existing offering to speed up the construction process by accelerating the construction scheduling and minimizing waiting time for workers when making efficient use of the elevator during the construction phase. The idea is to install an elevator in phases so that it can serve the construction site as early as possible. The lift jump, which lasts a few days, is always made when a suitable number of additional floors have been completed in the building. In the jump, the engine room of the elevator is disconnected from its temporary location and moved upward. At the same time, the elevator is inspected. After reinstallation, the elevator is as safe and reliable to use as the elevator of the finished building. During the jumps, installations are also made in the elevator shaft, which the finished elevator will eventually use. The JumpLift moves 2.5–4 m/s, while the speed of a traditional external construction hoist is 0.6–0.9 meters per second. The movement of employees between floors is significantly accelerated and working time is saved. Since the elevator is inside the building, bad weather does not stop it. Workers no longer need to compete for space with building materials, improving on-site logistics.

9.2 Industry Transformation Challenge

9.2.1 Background

The productivity stagnation problem in the construction industry has been debated for many years [1, 2]. According to the "Reinventing construction: A route to higher productivity" report, there is an opportunity to boost industry value by US$1.6 trillion [3]. The report highlights seven ways to improve the productivity of construction: reshape regulation, rewire contractual frameworks, rethink design and engineering processes, optimize procurement and supply chain management, improve on-site execution, infuse digital technology, and reskill the workforce. Improving on-site execution and infusing digital technology are the ones where KONE has played an active role in finding best practices together with the industry players. Digitalizing processes with on-site execution has been developed in an industry ecosystem [2]. The use of indoor positioning systems and sensors on all assets (labor, material, and equipment) has been developed to digitalize the construction site [1]. A common data platform is needed to share construction data with the key players on-site [4]. However, it is well known that the construction industry is highly conservative and fragmented, meaning changes happen slowly. The number of subcontractors is high, and work orders are divided into small pieces, leading to a situation where multiple companies need to adopt new practices. In addition, public authorities tend to slow down such developments as there is no single area of the construction industry that is standardized. There is a major transformation opportunity ahead in this area.

For decades, KONE has been a forerunner in using new technologies and operational innovations. Its new strategy includes a company-wide transformation for "digital + physical enterprise" that aims at having a future-proof technology infrastructure, building the capabilities to use data and analytics, and further developing the efficiency and resilience of its supply chain. The strategic importance of real-time data-driven digital services is rapidly growing in the elevator business as well as in the whole construction industry. To support the ongoing digital transformation, there have been several trials to develop new digital solutions and infrastructure for construction sites. For instance, ad hoc IoT mesh networks have been tested to enable data gathering from sensors and indoor positioning of wearable tags. Radio frequency identification readers (RFIDs) have been installed in building entrances and elevators to enable material flow tracking. On-premises computing servers have been tested for local data collection and processing with basic internet connectivity. APIs have been developed to enable developers to build proof-of-concepts (POC) and ideas for new digital innovations. Online equipment tracking was piloted to reduce the time spent searching for tools on-site and monitoring the tool usage rate in the back office. Selected tools were tagged so that their location could be monitored and shown to site personnel on a user interface. The usage rate was calculated from the time the tool was in motion. Situational awareness and analytics around material and people flow data were created by utilizing a real-time data platform to produce a situational picture of the construction site. Progress on tasks and potential issues on-site can be shared digitally in real-time with all project stakeholders to improve communication. Several key performance indicators (KPIs) were defined and calculated to measure progress and productivity.

9.2.2 Smart Construction Elevator

One interesting development area is the concept of KONE SiteFlow, where the building's elevator is installed early and used during the construction project to improve people and material flow on-site. The elevator acts as a data platform to optimize elevator control and improve transport capacity. Elevator usage data can be shared with other stakeholders on the construction site to improve productivity and share a common situational picture. The user interface was developed for end users to prioritize and monitor elevator use. The solution is still in the development phase. The solution includes a web application that encompasses functionalities that can be used on a building's construction site where KONE equipment is utilized. These functionalities include seeing from the user interface on which floor the elevator is and what its priority is, as well as allowing users to call elevators to a particular floor. The solution is being piloted, for example, at a construction site in Helsinki using the web application. Another version of the solution is also being piloted in London, where similar functionalities used in the web application are provided to the site through an API.

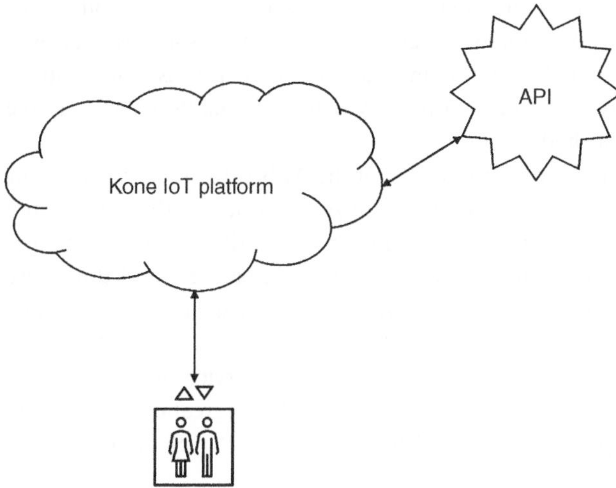

Figure 9.1 KONE IoT platform [5].

The solution is designed to utilize the same technology and operating model throughout the whole lifetime of the equipment – in installation, operation, and maintenance. The sensor data of the elevator are collected and transferred to the KONE IoT platform, where data can be turned into meaningful information and automated actions by utilizing data analytics and machine learning. In addition, the enriched sensor data can be transferred to customers, partners, or internal users through standard APIs. The same APIs can be used in exchanging information with the same stakeholders. The basic idea of the IoT platform is illustrated in Figure 9.1.

The IoT platform has been developed as a global solution over many years with a focus on full scalability and event-based architecture. The platform provides KONE with a core strategic capability to create and support digital solutions for all KONE businesses that are needed throughout the life cycle of the elevators. The platform provides standard APIs for developers and users that enhance the solution further [5].

9.2.3 The Importance of APIs in the Industry Ecosystem

KONE launched its first API for partners and customers in 2018, but it had been using APIs internally for a long time before that. In general, KONE strives to have generic APIs that enable modular, scalable, and easily customizable solutions for different use cases. An important API-enabled capability is to allow effective data sharing in the industry ecosystem – exchanging and refining data for different purposes. For instance, customers can utilize an API with their own mobile phone apps to make elevator calls digitally. A developer portal allows access to all APIs for customers and partners helping them to develop new use cases and applications. A technical support team can also easily help all portal users, as its implementation is well established.

There are different kinds of requirements for the APIs throughout the life cycle of an elevator. Construction firms want to utilize them during the construction project, whereas building owners and administrators utilize them for maintenance and facility management purposes. In addition, partners and third-party developers are increasingly interested in the APIs, for instance, for developing logistics services around the elevators. KONE has a Partner Ecosystem Program to boost close collaboration in creating new product and service innovations. The digitalization maturity of customers and partners differs a great deal, thus influencing which API approach to select. Therefore, the KONE SiteFlow solution allows both direct and indirect API usage.

The value chain of the API consists of five elements: business assets, API, developers, applications, and end users [6]. These elements describe how business assets are converted into value for the end user through the indirect API channel while generating value for each participant in the chain. Using these elements, the API value chain for the Kone Service Info API is described in Figure 9.2, which also includes a practical example with one of Kone's strategic service partners in the construction design and maintenance area.

The Kone Service Info API integrates information about service orders, callouts, and repairs into a user's building management system. The API provides construction engineering and facility management partners with an easy method to receive and inspect any maintenance service information and integrate it into chosen building management systems, giving a comprehensive dashboard and status of the maintenance activities.

KONE has seen value in opening its business capabilities to its customers and partners. This means that by combining KONE's equipment with modern technology, KONE can create new services and functionalities that are usable for customers digitally through some form of interface. By opening up its business capabilities, KONE has been able to create new digital products and services that relieve customer's pain points throughout the life of the building. For example, KONE has created a new solution that enables elevator calls to be made digitally on KONE's own Elevator Call mobile application or through KONE's Elevator Call API. Customers who opt to use Elevator Call API can integrate the functionalities offered into their own mobile application or into their building's lobby screen. This solution is available during the building's construction phase and operating phase. In both cases, this elevator call solution gives the customer an opportunity to control their equipment in a use case

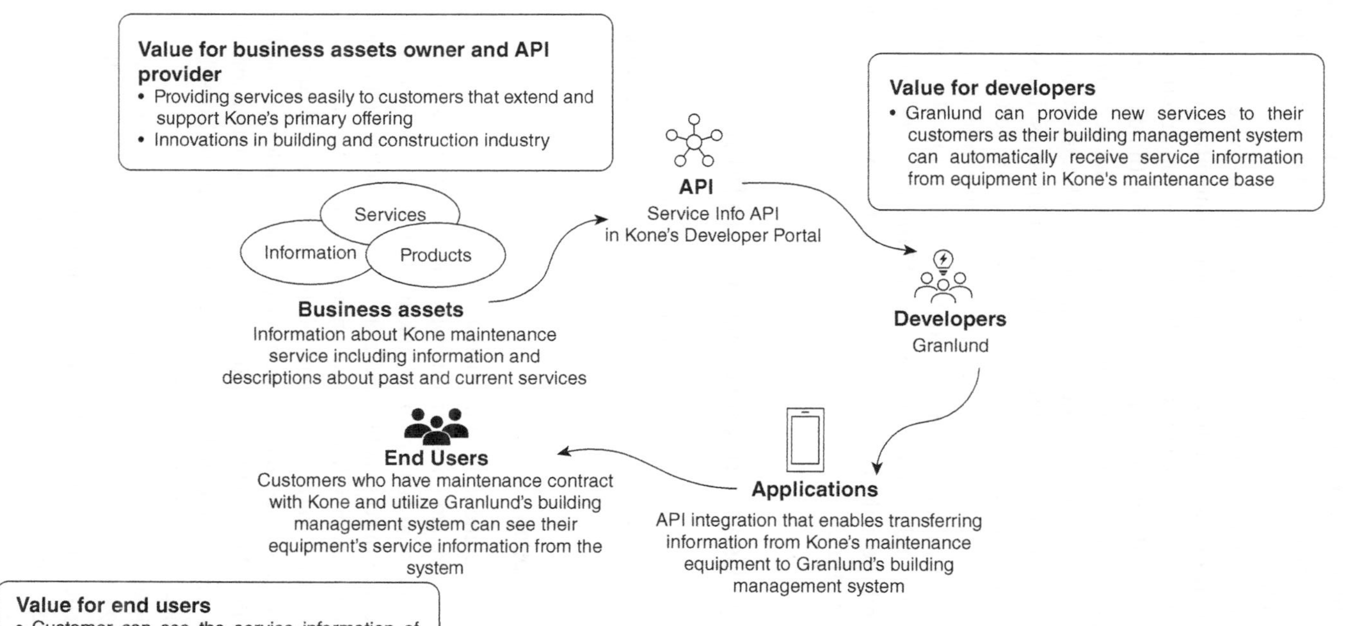

Value for business assets owner and API provider
• Providing services easily to customers that extend and support Kone's primary offering
• Innovations in building and construction industry
Services
Information
Products
Business assets
Information about Kone maintenance service including information and descriptions about past and current services
API
Service Info API in Kone's Developer Portal
Value for developers
• Granlund can provide new services to their customers as their building management system can automatically receive service information from equipment in Kone's maintenance base
Developers
Granlund
Applications
API integration that enables transferring information from Kone's maintenance equipment to Granlund's building management system
End Users
Customers who have maintenance contract with Kone and utilize Granlund's building management system can see their equipment's service information from the system
Value for end users
• Customer can see the service information of their equipment without creating end-to-end integrations
• Information is in the same system that is used for tracking other building related issues
• Manual workload of the customer decreases

Figure 9.2 API value chain for Kone Service Info API [5].

that is important to them. During the building's construction phase, the solution can help construction companies move their materials and tools in a manner that eases flow at the site.

During the building's construction, the APIs and digital services resulting from them can generate operational benefits through optimization of logistics that may lead to significant cost savings [5]. The faster the construction company finishes a building, the more profitable it is for them. In addition, construction companies might have to pay penalty fees if they deliver a building late. Construction projects are dependent on smooth and reliable logistics. Thus, by combining KONE's equipment with APIs and logistics, logistics at the site can be enhanced based on data. KONE aims to create solutions that enable understanding of the flows of vertical movement while enhancing them and giving customers opportunities to control them. In practice, queueing to use the elevator could be reduced as APIs could transfer data about possible bottlenecks, and the right people, tools, and materials would be at the right place at the right time.

9.3 Data-Driven Use Case: Construction Site Pilot

KONE SiteFlow is our use case to describe how 5G technology can be applied to share and utilize real-time data between stakeholders to improve the logistics of a construction site.

9.3.1 Use Case Description

Real-time situational awareness on construction sites is seen as one potential solution to improve construction work productivity and reduce the risk of project delays. A common situational picture for all stakeholders improves coordination of work, makes wasted resources visible, and enables a fast reaction to problems. With digital technologies, such as indoor positioning, sensors, data analytics, and connectivity, it is possible to gather data about people, materials, and equipment status on construction sites. These data are transferred using wireless connectivity to the cloud and analyzed to produce valuable situational information for stakeholders. Information can be presented using mobile applications, web dashboards, on-site displays, projectors, or even XR glasses.

A prototype of this was developed for a high-rise building construction site in Helsinki in 2022–2023. Development was carried out iteratively in close collaboration with the construction company. Alternative solutions were tested, evaluated, and prioritized throughout the construction project based on various stakeholder feedback.

9.3.2 Data Collection

Elevators used on a construction site provide crucial data to situational awareness systems about trips taken to transfer people and materials. Elevators can also be equipped with additional sensors to identify details of people and material movement. Elevators are logistics hubs that most things pass through at some point, especially in high-rise building construction, where the use of stairs is not practical. Elevators are also convenient places to undertake electronic tracking, as they are one of the few places in a hectic construction site where items remain still long enough to be detected.

KONE prototyped an elevator that could identify the contents of the elevator car. Identification was achieved by combining data from two sources: (1) the data produced by the elevator system about trips completed and floors visited, and (2) additional sensor equipment installed in the elevator car and landings to identify people and materials tagged with ID tags. RFID technology was used that enabled remote detection of tags up to 5 m without any user actions required. All site workers (close to 1000 people), some equipment, and materials were tagged with inexpensive sticker tags, as shown in Figure 9.3.

Elevator data and RFID tag data were provided by respective systems over APIs, as shown in Figure 9.4. The new service connected to these APIs and recorded all elevator movements together with people and movement data. Data were combined to produce information about the last seen location of people and items and a history of trips taken, i.e. a situational picture of the site.

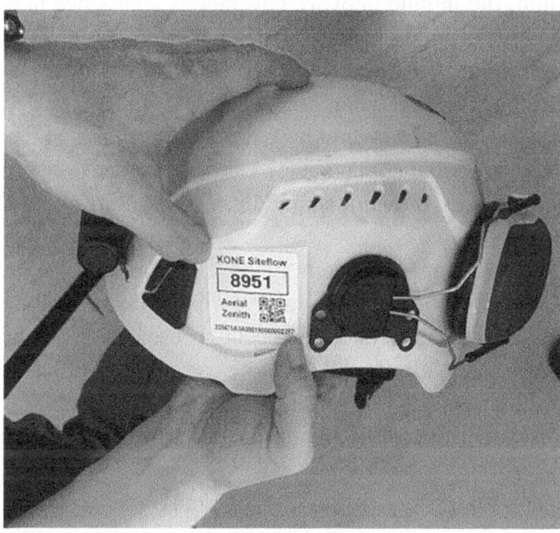

Figure 9.3 Worker helmets tagged with RFID tags.

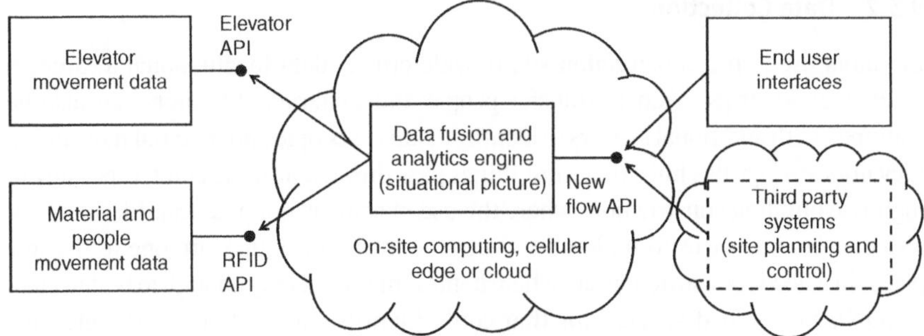

Figure 9.4 Simplified system architecture and data flow of the solution.

The new service provided this information as a new "Flow" API, enabling various user interface solutions to be connected. In the first experiments, the user interfaces were a mobile application and a desktop web dashboard. The API could also be opened for third-party access, allowing system integration with other construction productivity tools such as digital logistics planning and control solutions used on site.

9.3.3 Data Connectivity

Data connectivity on construction sites is typically handled using mobile networks. The building's fixed networks become available only in the latter phases of the construction project. Often, the mobile network coverage is poor since new buildings are built in areas where cellular coverage has not been needed before. Cellular network (4G/5G) capacity may not be adequate, or base station antennas may be directed to serve other areas. Even when buildings are built in crowded city centers, the newly built highest floors may not have coverage as the antennas are directed down, not up. An additional challenge for elevators is that they are built in steel-reinforced concrete shafts, often in the middle of the building, where cellular reception is the poorest.

Figure 9.5 shows the RFID antenna placement in an elevator car and on landings. To overcome the challenges of poor cellular coverage, a wired data connection was provided over the elevator system to a cellular modem on top of the elevator shaft for better coverage. This also allowed the use of a cost-efficient single connectivity hub for all data instead of connecting them all individually. The downside of this is the extra wiring needed, which increases the installation cost and complexity.

To improve connection reliability further, additional antennas were installed. These provided more gain and better directional capability and significantly helped connectivity in corners. Also, it is important to measure and choose the operator and bands with the best coverage. Some cellular modems enable the use

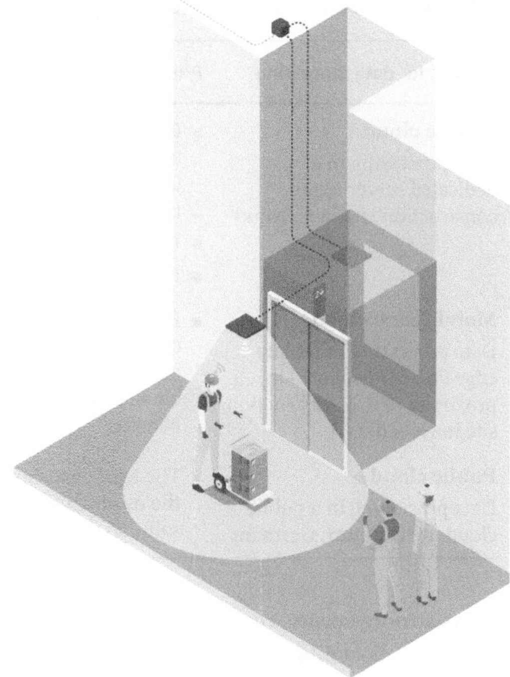

Figure 9.5 RFID readers collecting data from workers and material location in an elevator car and on landings.

of multiple operator SIM cards, and the network can be switched based on congestion. On the largest construction sites, with many device installations and other needs for data connectivity, temporary private networks, such as Nokia Digital Automation Cloud, can be considered.

9.3.4 Data Processing

All software for the pilot use case was designed to run in Docker containers based on cloud technology, allowing data processing to be easily moved to run either in the public cloud, 4G/5G mobile edge, or locally on-site in a dedicated server (private cloud) in the elevator shaft. All three architecture alternatives were tried. A comparison of the solutions is shown in Table 9.1.

9.4 Benefits of 5G

In the elevator industry, one of the pioneering data-driven innovations is KONE SiteFlow, which provides a construction ecosystem with a common data platform to manage information and material flows during a construction project. The concept has been further developed with 5G standalone technology that

Table 9.1 Comparison of data processing architecture alternatives.

Options for data processing	Pros	Cons
Private cloud Data processing in a dedicated server at the construction site in Helsinki	• Connectivity challenges are reduced as less data is transferred out from the site for processing • Lower latency • Perceived better security	• Hardware needed on site (higher cost, more complex installation, and reduced reliability in dusty and warm environments)
Mobile edge cloud Data processing in mobile edge (4G/5G network) in proximity to the construction site in Helsinki	• Less hardware is needed on-site while still experiencing some on-site hardware benefits	• Local unique contracts with operators are needed (poor scalability for global market)
Public cloud Data processing in a public cloud in Frankfurt, Germany	• The least expensive and the most scalable solution	• Higher latency in some (real-time) use cases • Perceived security risks

allows a more secure, real-time connectivity and computing platform. The basic idea is to collect different types of sensor data from elevators and, by using cloud-based ML/AI applications, turn data in real-time into meaningful information, enabling improved optimization of material flows on the site. Connection reliability and resilience are top priorities for the utilization of 5G. Another important aspect is the development of the concept with new 5G features that allow more flexibility and real-time capabilities to build new digital services and applications, such as video analytics, for the benefit of all ecosystem players. By using elevators, the KONE SiteFlow analytics service aims at improving the safety and efficiency of operations on the construction site as well as building new applications to optimize the material flow. The enhanced concept is illustrated in Figure 9.6.

An online 360° camera system attached to the elevator cars collects video streams and pictures inside an elevator. In addition to the video stream, all types of traditional sensor data from machines, tools, workers, etc. can be combined too. The collected data are wirelessly transferred using 5G standalone technology that allows a separate network slice, creating an isolated end-to-end connection capacity for the application. The 5G base station utilizes the standard 3.5 GHz frequency spectrum and is connected to a mobile edge cloud that is physically close to the construction site. With 5G technology, the user's plane data can be safely analyzed and treated on the mobile edge enabling low-latency connections back to the site. By using ML/AI-enabled data analytics applications, the user data can produce information on the occupancy rate of the elevator's floor area, the cubic volume of materials loaded inside the elevator car, the duration of the elevator's loading and

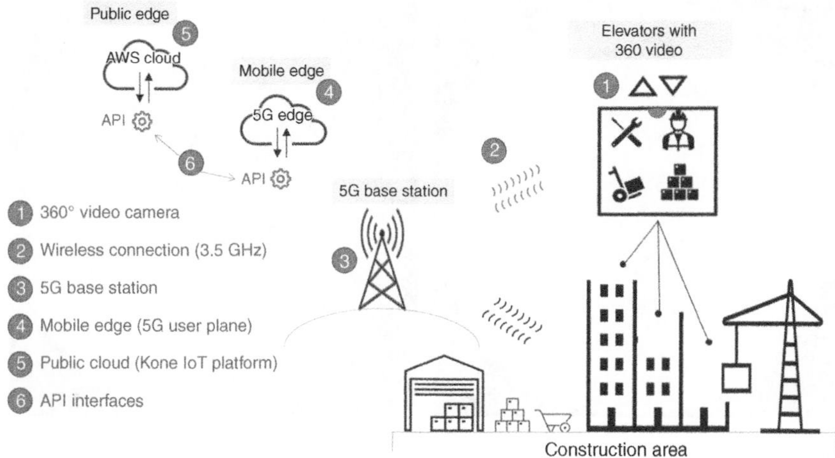

Figure 9.6 KONE SiteFlow with 5G technology.

unloading events, the identification of unnecessary stops in the elevator, and the number of elevator users. The service identifies humans but does not identify individuals. The anonymous data generated by the analytics are combined with the data produced by the elevator system. The information in the service is used to develop the operation of elevators and the efficiency of transportation. Running applications in a mobile edge, practically, requires the use of cloud-native containers that are typically based on containerization technology such as Docker or Kubernetes.

Another option is to transfer data to a public AWS cloud, where the application was originally developed as a cloud-native container to be easily deployed, managed, and scaled. Running KONE SiteFlow applications in the public cloud would be the normal way of working – when the latency requirements allow it. The use of standard APIs is an essential part of the solution, as it is often integrated with other stakeholders' applications and services on the construction site.

Figure 9.7 shows how the video stream can be analyzed in real-time by using a separate analytics software in the cloud. The analysis can contain different kinds of aspects needed in the KONE SiteFlow concept, for instance, the maximal use of elevator floor area, occupancy rate, elevator car internal cubic volume, movements of building materials, safety risks, identification of dangerous materials and tools, etc. Development is continuing to improve the optimization of material and people flow in the elevator.

There are various benefits of 5G technology – especially for applications where video streaming is utilized. These include when visual confirmation between the physical and digital worlds is needed, when the customer needs to know whether

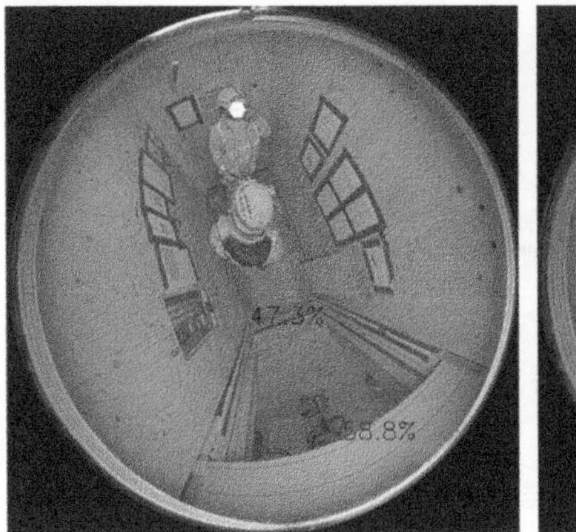
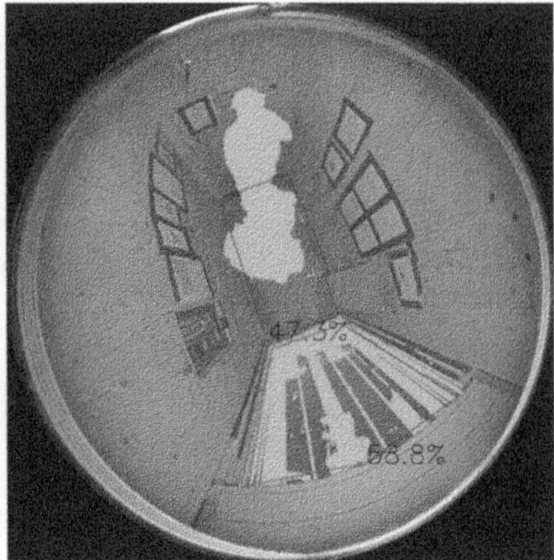

Figure 9.7 SiteFlow video analytics.

the elevator is operating, when a large amount of data is transferred, and when there are multiple sensors to be integrated. A basic 5G connectivity does not on its own directly create value, but mobile edge computing and new 5G features (such as slicing, IoT capacity, lower latency, and uplink/downlink bandwidth capacity) make it possible for new data-driven use cases. In the future, it is likely that these use cases will require higher bandwidth and greater energy efficiency. 5G enables data analytics to be executed in the cloud instead of on local devices, which reduces the cost of devices and helps with hardware and software release control. This also enables flexible algorithm development when hardware dependencies are minimized.

9.5 Future Opportunities

There are plenty of future opportunities to scale this type of real-time solution on construction sites. The use of video cameras in elevator cars and lobbies to monitor usage would extend the use case. Cameras in the elevator shaft can be utilized for elevator condition monitoring during construction phases, where live video streaming over networks is needed.

Video monitoring systems and their footage could be utilized as one data source for elevator traffic prioritization. Many sites already have good coverage of their lobby areas and landings with CCTV systems, and utilizing live footage in analytics would help to understand people's flow as part of the bigger picture. Creating local computing capacity and maintaining systems would not make a lot of sense, but analyzing already existing sources combined with mobile edge computing would be smart. The metaverse with AR/VR applications will require 5G technology in many use cases. For repair and maintenance, an online connection to a remote support center would be possible.

KONE is still in the experimental phase with the API ecosystem and platform, where "seeing, learning, and trying to find the ways to monetize this is taking place." KONE sees more potential for ecosystems and platforms in the future. Interviewees mention the possibility of selling data insights gained through KONE's equipment, KONE's partners finding new partners through KONE's ecosystem, as well as utilizing KONE's ecosystem's companies' sales departments for co-creation, all as possible benefits of future ecosystems and platforms.

A future model could be that all application development takes place in the AWS public cloud as a containerized solution, and its distribution could be executed "automatically" to the closest mobile edge for the users at the construction site. This setup would guarantee fast development through APIs and maximize the performance of applications (e.g. low latency, capacity, etc.).

References

1 O. Seppänen, J. Zhao, B. Olyaei, M. Noreikis, Y. Xiao, R. Jäntti, V. Singh and A. Peltokorpi, "Intelligent Construction Site (ICONS) Project Final Report," January 2019.

2 O. Seppänen, A. Peltokorpi, Y. Zheng, M. Masood, A. Aikala, J. Lehtovaara, M. Kiviniemi and R. Lavikka, "Digitalizing Construction Work Flow (DiCtion)," 2021.

3 F. Barbosa, "Reinventing Construction: A Route to Higher Productivity," McKinsey Global Institute, February 2017.

4 I. Lappalainen and S. Aromaa, "The emergence of a platform innovation ecosystem in smart construction," in Proceedings of ISPIM Connects Valencia – Reconnect, Rediscover, Reimagine, Valencia Spain, 29 November–1 December 2021.

5 E. Koivula, Building Competitive Advantage with API Strategy – Case Study of Established Enterprises, Espoo: Master's Thesis, Aalto University, 2023.

6 D. Jacobson, G. Brail and D. Woods, APIs: A Strategy Guide, O'Reilly Media, Inc., 2011.

10

Telecom Industry: Improving Energy Efficiency for Climate

CASE STUDY TEAM MEMBERS:

Roope Lahti: Aalto University
Janne Koistinen: Telia Finland
Eija Pitkänen: Telia Finland
Timo Saxen: Telia Finland
Jari Collin: Aalto University/Telia Finland

10.1 Introduction

The case company from the telecom industry is Telia Finland, which is part of the Telia Company group operating in the Nordic and Baltic regions. The purpose of the company is to reinvent better-connected living. Having millions of mobile, fixed voice, broadband, and TV customers, it provides the backbone of the digital society – connectivity – as well as innovative solutions that are vital in creating a better future.

The increasing energy consumption of ICT and related growing amounts of e-waste are contributing to global warming and unsustainable use of natural resources. On the other hand, digitalization can also be a key part of the solution and accelerate the transformation needed to change this course. Sustainability has become a core value and business driver for many companies, the telco industry included. This change is part of a shift in the mindsets of companies and investors that businesses must both adapt and combat climate change, and that the time for robust action is now.

5G Innovations for Industry Transformation: Data-Driven Use Cases, First Edition.
Jari Collin, Jarkko Pellikka, and Jyrki T.J. Penttinen.

Published 2024 by John Wiley & Sons, Inc.

Telia's business strategy captures both Telia's responsibility to address its negative carbon footprint robustly and business opportunities to help its customers reduce theirs. The company has committed to reach net zero carbon emissions by 2040. The target is aligned with the new SBTi Net-Zero Standard.

Telia Company's total value chain greenhouse gas (GHG) emissions in 2020 amounted to 1150 ktons of CO_2e (2018: 1175 ktons). The majority (99%) of emissions are generated in the value chain, whereas its own operations have a limited climate impact. This is due to the fact that Telia Company has, since 2020 used 100% renewable electricity in its operations.

Several research reports such as the Exponential Climate Action Roadmap, the GSMA Enablement Effect report, and a Telia/Accenture report on the circular economy show the importance of connectivity and digital solutions for societies to become low-carbon and circular. Telia Company is actively contributing as a supporting partner to the Exponential Climate Action Roadmap project, which has described 36 solutions to halve global CO_2 emissions by 2030 [1]. The project highlights the role of digital solutions as an enabler for many other industries, stating that globally, "The digital sector has the potential to directly reduce fossil fuel emissions 15% by 2030 and indirectly support a further reduction of 35% through influence of consumer and business decisions and systems transformation."

Digitalization is unquestionably one of the largest megatrends impacting not only the telecom industry but also the whole world – for the present and the future. The industry sector has enabled many traditional, existing services to move online, and many new digital services are constantly being created. For countless consumer applications, this transition to digital has already happened. Video streaming has all but wiped out DVD sales, online booking services have replaced travel agency offices, online collaboration tools have enabled remote working and schooling, etc. Many other examples exist on how data-driven online services have transformed legacy business models. Several research reports have shown the importance of connectivity and digital solutions for societies to become low-carbon and circular [1–3]. Since 2020, the company has tracked "enablement effects" for some of its products and services: the use of remote meeting services and some Internet of Things (IoT) solutions applied in buildings, transportation, and utilities enabled GHG emission reductions of 330,000 tons CO_2e. Apart from the carbon enablement effect, Telia also measures energy reductions enabled by IoT, which enabled savings of approximately 810 GWh of electricity in 2022.

Energy efficiency is built into the roadmaps of mobile technology – especially boosted by 5G technology. It is estimated that global mobile data traffic per smartphone will increase from 4.8 GB per month per user in 2018 to 45.9 GB in 2028 [4]. There are large variances between countries. In Finland, the world's most data-hungry country, mobile network data volume was already 55.9 GB per month per person in 2021 [5]. Even though this digital transformation has already progressed far in the consumer sector, the same cannot be said for the industrial sector. It is predicted that digitalization will bring on the fourth industrial revolution, which will lead to massive gains in production efficiency. At the heart of this transformation will be intelligent automation and optimization driven by massive amounts of data gathered from processes and the environment with IoT sensors and other M2M devices, in addition to various digital tools that make human labor more productive and meaningful. In fact, Cisco predicts that the fastest-growing device type in mobile networks in the coming years will be M2M devices [6]. Much of this industrial revolution will be powered by 5G, which introduces many new verticals, especially in low-latency, high-bandwidth, and ultra-reliability use cases. According to PwC, 5G will generate 13.2 trillion dollars in economic value by 2035 while creating 22.3 million jobs. Of this 13.2 trillion-dollar additional value, 4.7 trillion is expected to be created in the manufacturing industry alone [7].

This chapter first introduces the challenge that the telecom industry faces with increasing energy consumption and then introduces data-driven methods for how this problem has been and could be combatted in the future. 5G is then examined as a driver for these solutions with an overview of new functionalities and two case studies: one on 5G networks and one on edge computing, giving the first glimpse of what kind of energy savings could be achieved with these new technologies. Section 10.5 then looks further into the future to give an idea of what is possible as technologies advance even further and become more mature. Finally, a summary with the most important takeaways is given in Section 10.6.

10.2 Industry Transformation Challenge

As mentioned in the introduction, the two largest trends in the telecom industry today are increasing data volumes and pursuing sustainability. These two trends create a major challenge for the industry, namely increasing energy consumption and how it can be mitigated. Energy costs have always been one of the largest operational expenditures in the telecom industry, but the recent energy crisis in Europe has only made the situation more dire. These rapid increases are likely to remain only temporary, but the situation has shown the vulnerability of the energy market. It is likely that some upcoming crises will have a similar effect on the energy market, so companies cannot build their business in the hope of low and stable energy prices. It is also clear that neither the telecom industry nor the environment can cope with an exponential increase in energy consumption brought about by an expected exponential increase in data volume within a decade [4], and thus massive improvements in energy efficiency will be required now and in the future. Mobile networks are, by far, the most energy-hungry assets a network operator has, and, within a mobile network, base stations account for around 80% of all energy consumption. In fact, an average base station in an urban area consumes over 3 kW of power, which is equal to around 15 desktop computers. Therefore, improving base station energy efficiency should be at the heart of telecom industry's development.

In the past, energy-saving techniques have focused mainly on device component optimization. For example, LTE introduced a technique called discontinuous transmission (DTX), which can create large energy savings by switching off transmitting components when they are not needed. These device optimizations will continue to be important in the future, but with such a massive increase in data volumes and the number of connected devices, they will not be enough on their own. In addition, industrial use may introduce large-scale constant loads, e.g. video streaming from field cameras that generate continuous, uninterrupted uplink loads which make many traditional energy-saving methods unfeasible.

In addition to a large ongoing transformation in the telecom industry, an arguably even larger transformation is ongoing in the energy industry. With the same goal for increased sustainability, energy companies are heavily investing in renewable energy. A large part of this renewable capacity will be wind and solar power, which are volatile by nature, and the production volume can change rapidly. In addition, moving away from traditional massive steam turbines also reduces electricity grid momentum. These changes create two important changes in the market. First, energy prices become more unstable and can change immensely, even on an hourly basis. Second, as grid momentum is reduced, compensatory solutions are needed to keep grid frequency stable. One of the leading solutions to combat this problem is large-scale energy storage that can be used to stabilize the grid when needed [8], but large-scale deployment would require massive investments. Telco is in a unique position to assist in this transformation, as it already possesses a large, decentralized energy storage system in the form of base station backup batteries. With some modifications and investments, the backup system could serve both of these use cases. While total energy usage is not reduced through these solutions, they will promote the transformation to renewable energy sources and stabilize electricity prices.

It is therefore clear that we are in the midst of a massive change in both the telecommunications sector and the energy sector, where new innovative solutions are desperately needed. Utilization of real-time data has, thus far, been a mostly untapped area for energy efficiency gains. This is a pity, since in this age of abundant data, all processes should be automated as far as possible, and all business decisions should be based on real data rather than gut feelings. Thankfully, solutions are already available, and more are constantly being developed. These solutions can roughly be split into two categories: automated machine learning (ML) models that utilize real-time data to make independent decisions without human input, and online dashboards that gather data and present it in a meaningful way so that manual action can be taken based on the data. The first option is able to improve energy efficiency in operational systems constantly, and the second assists in making informed decisions on new investments and projects. As we will see later, an important driver in both of these categories is the increasing availability of analytical models that can be used to draw useful conclusions from the data. After all, data on its own has no value; it is only useful when it is analyzed and then used to make changes in operations. The industry transformation challenge in the telecom industry can thus be formulated in the following way: How can real-time data from system components be used to improve the energy efficiency of 5G networks, edge computing, and customer industrial processes? This study approaches the transformation challenge in two parts. First, a survey of current practices in mobile network design and operation is carried out to find the low-hanging fruits in data-driven solutions. Then, the energy efficiency of both 5G and edge computing is measured to determine how they compare on their own to competing technologies.

10.3 Data-Driven Use Cases

Telecom operators have access to large amounts of interesting data that has long been underutilized. A great deal of these data are collected from mobile networks. Simple to operate, mobile networks require complex information on user movement and the state of base stations. If we consider how just about everyone owns at least one device with a SIM card and how many base stations are needed to cover an entire country, the amount of data available is staggering. The number of data points is only going to increase in the future with denser base station deployments and M2M devices. Only recently, these data have started to be utilized in commercial applications by building products around them. One of the most attractive and obvious applications is anonymized data from crowd movements and commutes, which can be used by various industries including public transport operators and the retail industry. 5G will only expand upon this offering with 1-m accurate location-based services. In addition to offering anonymized data as a commercial product, endless opportunities exist for using this data to improve the company's own operations.

One of these opportunities is improving energy efficiency. Generally, telecom operators cannot decide how network equipment is designed or programmed, but they can decide how it is used and where it is placed. Choosing the correct device model, optimal position, and configuration is, in fact, a very complex process where energy efficiency is usually overshadowed by factors such as higher peak data rates. In some situations, such as in heavily populated areas, pursuing these high data rates makes sense, but are they truly needed in sparsely populated areas?

In some situations, the decision to opt for lower-power equipment is obvious for a network planner, but not always. Networks are also constantly becoming more complex with new technologies, especially with cell densification and the use of higher frequencies. With previous network generations, a single BS could cover a very large area since frequencies were lower, but with increasingly high frequencies and the demand for increased data rates, cell coverage is becoming smaller and smaller. In addition, recently there has been increasing discussion on microcells that would cover small indoor areas similar to traditional Wi-Fi access points. Network planning is, thus, becoming increasingly complex and unfeasible with current resources.

In a situation like this, ML, combined with data collected from the mobile network for various data points such as base station energy consumption, data volumes, and crowd movements, could be utilized to make this process easier. This would also give network planners valuable feedback on their work which could otherwise go unnoticed. In its most basic form, the system would serve as an online dashboard showing historical data on network performance and typical energy consumption and data volumes for different device models. This would already make network planning more informed, but if the system is fed with additional data and more powerful algorithms, it could be even more helpful. A network planner could select a position on a map where a new base station is needed, after which an algorithm would analyze previous traffic and surrounding base stations in the area and suggest the best configuration for performance and energy efficiency. With further algorithm training, the system could begin to find these opportunities proactively by monitoring the network. It could give a list of locations it has found where the network is not functioning optimally due to inefficient device configurations and then suggest alternative options that would perform better.

Data-driven energy optimization opportunities can also be found in day-to-day mobile network operations. Data volumes experience huge fluctuations between day and night, and base stations are usually intended to meet peak-hour demand. Therefore, base stations are underutilized for most of the day. As discussed before, modern networks usually consist of many layers, where small microcells provide local high-bandwidth network access and macrocells provide coverage with limited throughput. Therefore, during idle times, some high-capacity small cells could be powered down to save energy, and coverage would still be maintained by larger cells. Due to a lack of high-quality data, traditional approaches have consisted of statically timed shutdowns at certain frequencies, which is far from an ideal approach. As with network planning, an algorithm fed with data from the whole radio access network (RAN) could be used to identify these potential energy savings and then act upon them. A truly data-driven solution would be much more opportunistic, so it would be likely to find shorter intervals for component shutdowns during the day as well, once the system became able to model usual network behavior accurately. The potential for industrial environments is also considerable. Even though data volume fluctuations are not as great as in consumer networks, as industrial processes take place at night as well, fluctuations still exist. For example, the work hours in a factory environment are very consistent, as breaks are taken, and shifts change at the same times from day to day. An algorithm would, therefore, easily be able to identify the times when a production line is active and when it is not and power off surrounding networking equipment accordingly.

Real-time data will also play an important role as the amount of renewable energy increases across the electricity grid. Solar and wind power's impact on grid stability has been identified as a key issue preventing full-scale adaptation, and no singular solution to fix the issue has yet been found. Instead, it seems that the solution must be found in changes to how electricity is consumed. Traditionally, electricity production has always been adjusted to meet electricity demand, which has been straightforward as production capacity has always been available. With solar and wind energy, this practice needs a major overhaul, as solar energy is only available during the day and wind energy when it is windy. Therefore, it seems that in the future, demand must instead follow production. Automation and efficient energy storage have been suggested as the most promising solutions to this problem.

With automation, energy usage can be adjusted to occur only when electricity prices are low. For example, heating or heavy computing could be timed to occur only when energy production is high and prices are low. Naturally, this solution would require real-time energy prices to be communicated to the automation systems. Here, the telecom industry can create value for both itself and others. They can time their own operations in data centers to consume the most energy when prices are lowest and also encourage customers to do the same by providing access to real-time data and encouraging the moving of computing and data transfers to off-hours by product pricing. For example, data and computing usage could be made cheaper when energy production is high and usage is low. For private consumers, these changes in prices could be communicated through a mobile app, and for companies through an API that would encourage system automation. With intelligent pricing, telcos could meaningfully change usage patterns to meet production better and thus promote renewable energy investments.

Efficient energy storage has long been a problem, and a scalable solution has yet to be found. A part of the solution could be found in mobile networks. Base stations are usually equipped with a battery system that is used in case of power outages. In addition to serving their primary purpose, these battery systems could be an important asset in the future to act as a frequency reserve for the electricity grid. While a single base station does not have meaningful capacity on a national grid scale, if they are used as a centrally controlled system, their combined capacity can be very large. Grid companies are usually willing to pay for such a reserve, creating extra income from infrastructure that network operators already have. If the battery systems are large enough, they could also be used for peak shifting, which means using battery power when energy prices are highest and charging the batteries when it is lowest. For efficient peak shifting, the optimal times should be computed based on electricity price trends and predictions and base station energy consumption. Here, edge computing offers an attractive solution, as it can aggregate data from a central system with energy data and local base stations.

10.4 Benefits from 5G

While the last section focused on data-driven use cases that are possible today, this section looks at the impact that 5G specifically will have on energy efficiency and availability and processing of data. 5G energy efficiency will also be compared to 4G, and edge computing to on-premises computing, in a like-for-like case study to assess the energy efficiency of the technologies themselves. After all, data-driven solutions to improve energy efficiency are of no use if the underlying technology consumes more energy than it can save.

As we have seen, 5G has been designed from the beginning to serve many new industrial verticals, many of which can be very data-hungry. This creates a problem for telecom companies since mobile network energy consumption is already one of the largest operational expenditures that they face. This issue will only grow more severe with greater data volumes if energy efficiency remains the same. Luckily, 5G introduces many new methods for energy savings, even targeting a 100-fold improvement compared to 4G. Most of these new energy-saving features are targeted toward RAN, since it consumes around 80% of all electricity consumed in a mobile network, including user equipment radio modules. Energy savings in RAN boil down to two basic principles: high spectral efficiency and powering down equipment as much as possible. Higher spectral efficiency means that a higher number of bits can fit in 1 Hz of bandwidth. Spectral efficiency is highly dependent on signal conditions such as the signal to interference and noise ratio (SINR), but in ideal conditions, 4G is able to transmit 15 b/Hz in downlink mode and 3 b/Hz in uplink mode, whereas 5G is able to transmit 24 b/Hz in both downlink and uplink modes. As we can see, the difference is especially large in uplink mode, where 5G has a spectral efficiency of almost an order of magnitude larger than 4G. This improvement serves industrial use cases especially well, as many of them have symmetrical transmission patterns rather than a traditional downlink-oriented pattern of consumer use cases. As energy efficiency in mobile networks is commonly defined as the number of transmitted bits per energy consumed, a higher spectral efficiency also leads to a higher energy efficiency since more data can be transmitted using the same channel resources.

The second method for improving energy efficiency in RAN involves various sleep modes, in particular DTX. DTX allows transmitter components to be powered down when there is no data to transmit for durations as short as a millisecond. In 4G, DTX is severely limited by various mandatory control signals that must be sent periodically even when there is no traffic, which limits the amount of time the transmitter can remain in sleep mode. Many of these limitations have been removed from 5G. In addition, 5G supports sub-mmWave bandwidths of up to 100 MHz compared to 20 MHz in 4G. This means that combined with a higher spectral efficiency, data can be transmitted much faster over 5G. This, in turn, means that the transmitter can stay in sleep mode for longer durations, improving energy efficiency.

What do these improvements mean in the real world then? A first-of-its-kind case study that studied energy efficiency of production-use 4G and 5G base stations found great differences between the technologies [9, 10]. The study included measurements gathered over a period of one month from over a thousand base stations operating at 2–3 GHz. In the study, 5G was found to be much more energy efficient compared to 4G, especially when it came to load-based energy consumption. While an idle 5G radio unit consumes around twice the amount of energy that a 4G radio unit does, this difference quickly becomes insignificant with high and consistent data volumes in industrial environments. When looking at load-based energy consumption, on the other hand, it is much more in 5G's favor. While high band 4G consumes around 0.015 kWh/GB, 5G with a similar carrier bandwidth only consumes 0.004 kWh/GB. To visualize the difference in idle and load-based energy consumption, Figure 10.1 shows a comparison between a 4G and 5G base station energy efficiency. As can be seen, there is a clear limit in data volume where 5G becomes more energy efficient than 4G. In this case study, the limit was found to be around 40 GB/h. At first, this may sound high but when considering an industrial environment, this limit can be achieved easily. For example, two 4K video streams already produce enough data to make 5G more energy efficient than 4G, and when we consider the vision of an Industry 4.0 factory, each worker with an AR headset or remotely controlled robot can easily achieve this data volume on their own. The exact energy consumption numbers are specific only to the device models and software that were used in the case study, but the same pattern is likely to be found in other device models as well.

To put these numbers into perspective, let's consider them using an example. In the example, we consider an industrial environment where cameras are used to monitor various aspects of the environment, such as general safety and production quality. Twenty 4K cameras are used to provide high-quality, easy-to-interpret footage, and since the production lines are operational around the clock, the video

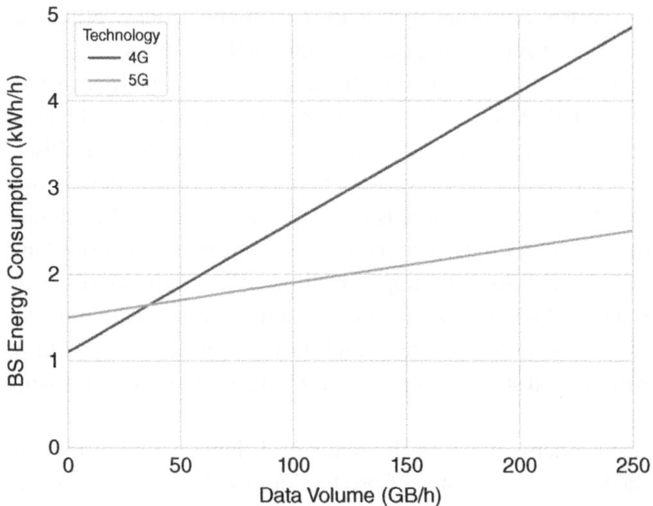

Figure 10.1 4G versus 5G energy efficiency [9].

streams are also continuous. A single 4K video stream is around 18 GB per hour and, based on the case study, the idle consumption of a 4G base station is 1.1 kW and a 5G base station 1.5 kW. Based on these numbers, we can make approximations on possible energy savings over a period of one month. The following table shows a comparison between 4G and 5G [9]. It is important to note that based on the results, the capacity of the 4G base station was not enough to transfer 360 GB per hour and thus two base stations are required to transfer the data.

	4G	5G
Base consumption	2 × 1.1 kWh	1.5 kWh
Load-based consumption	360 GB × 0.015 kWh/GB	360 GB × 0.004 kWh/GB
Total consumption per hour	7.6 kWh	2.94 kWh
Num. of hours in a month	720 h	720 h
Total energy consumption	5472 kWh	2117 kWh

As we can see in this example, 5G is around 2.5 times more energy efficient than 4G [9]. These are already impressive numbers, and they will likely grow even further apart with future hardware and software. 4G is already at the end of its 10-year development cycle, but 5G development is only at its beginning. It is likely that future technical specifications and hardware models will provide large improvements in energy efficiency with techniques such as deeper sleep modes.

Many new technologies made possible by the new 5G architecture will also play a key role in making processes more energy efficient and making even further strides with the assistance of real-time data. Edge computing has often been called the enabler of many new industry verticals, such as low-latency remote machinery control. Much of this demand for low-latency computing is also driven by battery-constrained devices that would benefit from computation offloading. With these devices, edge computing could extend device battery life considerably by handling all heavy computation tasks while, at the same time, keeping latency almost as low as carrying out computation on the device. To realize this goal fully, edge computing servers need to be placed as close as possible to where they are needed, even in conjunction with base stations. This new paradigm can provide the flexibility of cloud computing while still keeping latencies low. Customers will be able to start up new computing instances quickly and then dynamically scale them as they see fit. This is possible because the same platform is shared by all surrounding customers, so even if one's demand changes, the average demand remains stable. After initial processing on the edge, data can still be sent to the cloud for storage, but as most of the processing is already completed, data volumes are much lower. For example, analysis of a continuous video stream could be computed at the edge, and only possible security breaches could be sent to the cloud for storage.

As edge computing will especially serve low-latency applications, it is likely that most computing will be moved to the edge from on-premises servers. Here, edge computing offers major improvements in efficiency, as it is able to pool computing from multiple nearby facilities to the same server. Many current industrial use cases only need a modest amount of computing, so a traditional approach has been to install a small industrial computer in conjunction with the system it serves. For example, when looking for manufacturing defects on a production line, it is common to have a small computer on the production line that runs an algorithm to detect defects. This is far from an ideal solution, as each computer must run the operating system and other mandatory software components separately for each computing load. When these processes are all pooled at the edge of the network, as shown in Figure 10.2, the process becomes much more efficient, since larger servers can run tens of these workloads on a shared operating system, thus reducing overheads considerably. The edge computing provider therefore actually increases their own energy consumption but, in fact, reduces energy consumption when viewing the system as a whole.

As the number of devices utilizing these edge servers increases, optimal server placement quickly becomes an issue. Here, real-time usage data for the 5G network could be used to remedy the situation. Data processed on an edge server never leaves the 5G network, so it is easy to map where the data is coming from and where additional servers would be needed to provide an optimal experience

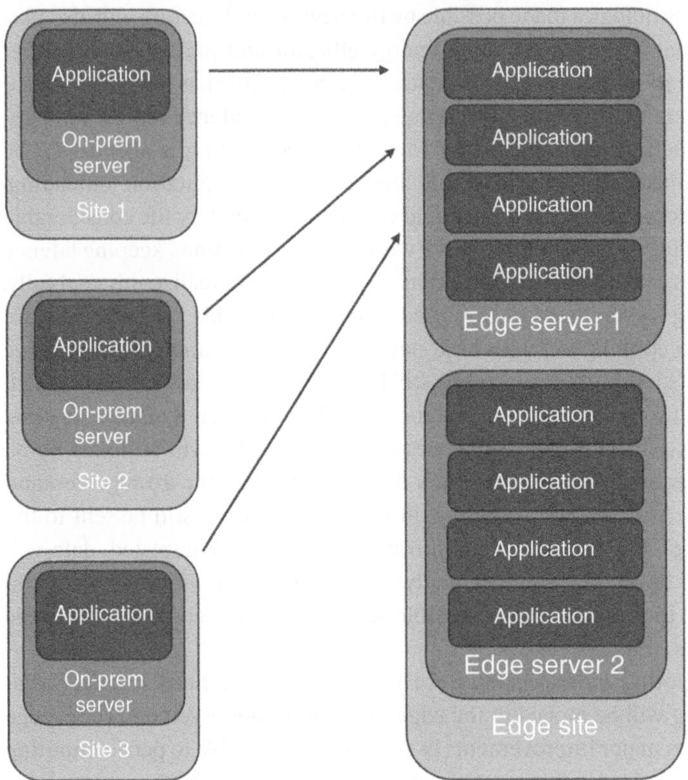

Figure 10.2 On-premises server versus edge server architecture [9].

for end-users. Furthermore, if 5G network slicing is used to divide devices into categories based on their latency tolerance, this data could be used to optimize the system further and ensure that these latency limits are met on all devices. In addition to reducing latency, nearby placement of computing can also reduce energy consumption. After the data is transferred across the over-the-air interface from user equipment to the base station, any further transmissions usually occur over fiber optic links. While transmissions over fiber optics do not require nearly as much energy as over-the-air transmissions, they will become increasingly energy-hungry as data volumes increase by orders of magnitude. While the energy consumption of small form-factor pluggables (SFPs) and routers on their own does not notably increase with data volume, they do still have a finite capacity. Once this capacity is reached, the network operator must invest in new hardware, thus increasing energy consumption. As data volumes increase with new industrial verticals, the current cloud-driven architecture could become unworkable for the core network. Therefore, if computing is moved from data centers to where it is needed, data volumes across the network become much lower, and additional fiber network devices might not be needed.

In addition to optimizing the placement of edge computing servers, data plays an immense role in energy efficiency optimization for edge computing servers themselves as well. According to the edge computing vision, computing loads should be designed in a way that they can run on any edge server in the network and can be moved to another server without disruptions when needed. This is necessary to facilitate end-user mobility and handovers to neighboring base stations. This characteristic mobility could also be used to optimize server workloads. All server hardware has a load level that is most energy efficient for it due to nonlinearity in performance per watt at different load levels. This level changes between hardware but, to achieve maximum energy efficiency, this level should be pursued. With real-time data on server utilization, computing loads could be divided among servers in the same area so that as many servers as possible are on this load level. In addition, servers usually have an idle consumption that is considerable compared to load-based consumption. Therefore, in situations where computing load is low, remaining load should be moved to a set of servers, and the rest should be powered off to reduce energy consumption. Industrial use cases usually produce computing loads that are more stable than consumer use cases since production lines usually run around the clock. Some fluctuations still exist, as usually some production lines only operate during the day and even more only on weekdays. If these same edge servers are also used in consumer applications, the shifts in load levels between day and night could be substantial.

It has thus far become clear that analytical data volumes produced by modern mobile networks currently, and especially in the future, are so large that machine learning will become an increasingly important data analytics tool. Unfortunately, current solutions for data gathering are splintered into many different systems, as there is no clear standard implementation for data analytics in networks. These data silos are severely limiting data usability, as it is impossible for any algorithm to gain a complete image of the whole ecosystem. This problem has been foreseen by the 5G NR standard-developing organization 3GPP, and for this purpose, they have defined a novel network function called network data analytics function (NWDAF). Even though the technical standard has been specified and 5G networks have been deployed, NWDAF has not yet seen much functional development or deployment. Therefore, it is difficult to list all use cases for the function, but some have been identified by research and device manufacturers. The common factor among these newly envisioned use cases is bringing more value through data. In some use cases, data about device movement can be used to optimize factory operations further, and, in others, the data are used to optimize the network itself further. One major area of focus for these use cases will likely be improving network energy efficiency.

NWDAF can be subdivided into two separate functions. Model training logical function (MTLF), which is responsible for training ML models based on data, and analytic logical function (AnLF), which is responsible for responding to queries from other network functions and administrators, utilizing the trained ML models. The great benefit of MTLF is that it does not rely on a set of ML algorithms defined by 3GPP but is instead an open platform for operators to implement the algorithms they need. This opens up endless possibilities for new verticals, since the amount of available data is abundant. Network operators could even start selling client-specific analytics as a service, as a specific algorithm could be used for specific network slices that a certain customer uses.

AnLF will be a great improvement for any data-driven use case, since it provides a standardized and programmable interface for requesting and receiving data and analytics. Previously, data analytics have suffered from inaccessibility and dispersion of data so, with a single interface, many more applications will become feasible and easier to implement.

It is easy to see how in a 5G network, almost all use cases that have been listed in the previous data-driven use cases could be deployed on the NWDAF. One algorithm could monitor network performance and create suggestions where additional equipment would be needed, another algorithm could power off equipment when it is not needed, and so on. Therefore, it can be said that NWDAF will play a central role in any data-driven use case. In conjunction with edge computing, it also transitions mobile network providers from a traditional infrastructure role to a platform provider for all analytical needs that a customer might have, from movement analytics to trends in data usage.

10.5 Future Opportunities

All the opportunities discussed previously in this chapter are ones that are already implemented or could be implemented within the very near future, but what if we look at the later stages of 5G roll-out and even further? One of the most attractive opportunities is to expand upon the data analysis used at present to create a digital twin of the whole 5G network. Digital twins have been discussed in manufacturing and construction industries for some time, but digital twins for networks offer opportunities just as lucrative. A true 5G digital twin would include everything from the core network to the base stations; in addition to monitoring network traffic and performance, it would monitor physical properties such as temperature as well. Here, the low latency and high throughput of 5G networks will play a key role, since the amount of transferred data needed for an up-to-date digital twin can be immense. 5G also greatly expands upon the support for low-power IoT devices, which will be critical to provide accurate environmental metrics from the base stations.

With a digital twin, not only could network operators understand the current state of their network better, but they could also simulate new implementation plans first virtually before implementing them in the real world. This would also give much more power to ML-driven energy optimization algorithms discussed in the previous sections, since the algorithm could simulate a countless number of improvement ideas before suggesting the optimal solution. In addition to optimizing physical hardware choices and placement, the digital twin could also be used to optimize software configurations. Currently, factories are designed to be modular by nature, meaning that new production lines are added, old ones are removed, and existing ones are moved throughout the lifetime of the factory. In fact, this is one of the major reasons why 5G fits into a modern industrial environment in the first place. The challenge that this production line shuffling brings is that communications demand changes as well. It could be, for example, that a data-intensive production line where each worker wears an AR headset is moved to the other side of the factory, and a previously busy mmWave microcell becomes idle. This would unlikely be noticed by network administrators, which would lead to unnecessary energy wastage, since mmWave transceivers can be quite energy-hungry. Using a digital twin, this problem could be spotted at the planning phase. If this issue is ignored in the planning phase, an ML algorithm would detect this unnecessary consumer of electricity during deployment and switch the whole cell off until it is needed again. Digital twin networks have been included as a future opportunity, as there are still some problems that need to be resolved before large-scale deployments are possible. These problems include:

- Upkeep of the digital twin, meaning the transfer of data and computing it, must require less energy than the technology is able to save.
- Algorithm scalability in large deployments is still limited.

While nationwide digital twin networks are likely still to be years away, digital twins for private networks offer an attractive compromise due to their smaller scale. Private industrial 5G networks usually only service a single site, so the amount of networking equipment is manageable. Higher data volumes may also offer an opportunity for greater energy saving, making them pay for themselves faster. It is also important to note that digital twins can offer higher energy efficiency, such as closed-loop network automation, which will play a key role in investment viability. It is important to note that the telecom industry also plays a key role in building a digital twin for industrial environments themselves. Digital twins for factories present countless benefits in automation and system visibility, many of which share similarities with digital twin networks. The mobile network provider could, in fact, position them as a service provider for digital twin services, instead of just acting as a communication relay. Edge computing with an integrated NWDAF function would create a powerful platform for digital twins, as companies would have easy and low-latency access to them.

In fact, the future vision for edge computing where computing is integrated even on a base station scale brings endless opportunities for new computing paradigms. Only time will tell how effectively the new architecture can be leveraged to improve energy efficiency, but it is clear that ML will be at the heart of all solutions, as the systems would be far too complex to manage manually. The main problem that these ML algorithms will try to find an optimal solution to is optimal computation offloading. This task is not trivial due to the sheer number of devices. They will consist of everything from tiny IoT sensors all the way to large industrial robots, and with edge computing servers at every base station, there is also the possibility to offload computation to neighboring servers. In an ecosystem like this, the computation scheduling used can either make or break energy efficiency, since with good scheduling devices can always operate in their most energy-efficient state but with bad scheduling, they operate very inefficiently. Scheduling also has an immense impact on delay, since if all devices decide to transfer their data for computing simultaneously, there will be a large spike in computation delay. In previous research, energy-aware scheduling algorithms have been able to reduce edge server energy consumption by 20% and UE energy consumption by 10% in simulated conditions [11, 12]. As the number of edge computing deployments increases, these algorithms must be evaluated again in production environments to see if the simulation results are applicable, but it is clear that efficient offloading and scheduling will have a great impact on system energy efficiency.

If we move focus more from the server side to user equipment, the next great leap in energy efficiency will happen in low-power IoT sensors. Currently, these devices can run for years on small batteries, so energy consumption of a single sensor is not significant. The problem only arises when the number of these sensors grows massively, which is expected with the Industry 4.0 transition and digital twins in general. This offers an opportunity for telcos to offer gains in energy efficiency in both their customers' and their own operations without increasing their own energy consumption. Using radio frequency energy harvesting and backscatter communication, IoT sensor batteries could be eliminated entirely, and energy from base station radio transmissions could be used to power the devices instead. Most people are familiar with backscatter communication from RFID tags, where the reader sends a signal that the tag encodes data and reflects it back to the reader. Energy harvesting devices are a bit more complicated, since they can collect and store energy from wireless signals, which also makes on-device processing possible. This means that the device can be used for much more than just to send a static identifier when queried, but instead send back measurement data such as temperature or angular velocity. The great benefit of backscatter and energy harvesting is that not only does it remove batteries from sensors, but it also does not increase base station energy consumption either, since radio signals must be transmitted in any case for communication. Therefore, it is an untapped source of otherwise wasted energy that can be used to improve energy efficiency in low-power devices. The current problem with implementation is that the 5G standard does not yet support backscatter communications, and energy consumption of many IoT sensors is not yet at a low enough level to mean that energy harvesting could be considered as the primary energy source.

10.6 Managerial Implications

As we have seen, increasing data volume creates a large transformation challenge for the telecom industry but, thankfully, energy efficiency delivers a promising solution. Solutions can generally be divided into two areas: improving energy efficiency by enhancing device efficiency, and by utilizing real-time data in network optimization. As we have seen, 5G enables massive gains in both of these categories. The main findings from the case study are the following:

- With a high enough data volume, 5G is much more energy efficient than 4G. In industrial environments, this limit should be easily achievable, as deployments are usually dense.
- Edge computing offers notable improvements in energy efficiency for industrial applications that cannot be moved to the cloud due to latency and security limitations.
- Data-driven ML algorithms have a massive potential in network planning, both in improving performance and energy efficiency.
- 5G core function NWDAF will play a key role in aggregating and analyzing 5G network data. It is also a key piece in building a unified analytics platform for internal and external use.
- Digital twin networks offer an attractive future opportunity, as they will make network planning more accurate and help with network anomaly detection. Some challenges still need to be solved before large-scale deployments are possible.

It is clear that 5G, edge computing, and a data-driven approach to energy efficiency form an integral part of any company operating in the telco industry, and the first adopters of a truly data-driven solution will be likely to gain a massive advantage in the market. The identification of future possibilities enabled by 5G is still ongoing in the telecom industry, and most have likely not yet been identified.

References

1 J. Falk and O. Gaffney, "Exponential Roadmap," Stockholm, 2019.
2 GSMA, "The Enablement Effect – The Impact of Mobile Communications Technologies on Carbon Emission Reductions," London: GSMA Head Office, 2019.
3 L. Rheinbay, M. Lieder, A. B. Töndevold and A. Holst, "The Shift – The Role of Telcos in the Circular Economy," Telia and Accenture, 2021.
4 Ericsson, "Mobile Data Traffic Outlook," 2022. [Online]. Available: https://www.ericsson.com/en/reports-and-papers/mobility-report/dataforecasts/mobile-traffic-forecast. [Accessed 1 March 2023].
5 Traficom, "Finland has More Broadband Subscriptions than the Other Nordic and Baltic Countries," 3 10 2022. [Online]. Available: https://www.traficom.fi/en/news/finland-has-more-broadband-subscriptions-other-nordic-and-baltic-countries. [Accessed 3 October 2022].
6 Cisco, Cisco Annual Internet Report (2018–2023) White Paper, 2020.
7 PwC, The Impact of 5G: Creating New Value Aacross Industries and Society, World Economic Forum, 2020.
8 EU Commission, "Energy Storage," [Online]. Available: https://energy.ec.europa.eu/topics/research-and-technology/energy-storage_en. [Accessed 30 October 2022].
9 R. Lahti, Assessing 5G and multi-access edge computing energy efficiency for industrial applications, Espoo: Master's Thesis, Aalto University, 2023.
10 R. Lahti, J. Manner, P. Mähönen, J. Pellikka and J. Collin, "Energy Efficiency of Modern Cellular Networks," IEEE Communications Magazine (Submitted), 2023.
11 K. Zhang, Y. Mao, S. Leng and Q. Zhao, "Energy-efficient offloading for mobile edge computing in 5G heterogeneous networks," *IEEE Access*, vol. 4, pp. 5896–5907, 2016.
12 M. Sana, M. Merluzzi, M. Pietro and E. Strinati, "Energy efficient edge computing: when Lyapunov meets distributed reinforcement learning," *in 2021 IEEE International Conference on Communications Workshops (ICC Workshops)*, pp. 1–6, 14–23 June 2021.

11

Oil and Gas Industry: Improving Operations with 5G-Enabled Drones at a Refinery Area

CASE STUDY TEAM MEMBERS:

Maarten van der Laars: Aalto University
Visa Oksa: Neste
Janne Anttila: Neste
Jari Manninen: Neste
Jari Collin: Aalto University

11.1 Introduction

The case company for the oil and gas industry is a Finnish industrial enterprise Neste that produces sustainable aviation fuel, renewable diesel, and renewable feedstock solutions. It is the world's largest producer of renewable diesel and renewable jet fuel refined from waste and residues, introducing renewable solutions to the polymers and chemicals industries. The company develops chemical recycling technologies and capacities to combat plastic waste challenges, with a purpose to create a healthier planet for our children. This major goal drives Neste to search for new ways to reduce amounts of greenhouse gases released into the atmosphere and develop innovative circular solutions for reusing carbon again and again. The company aims to become a global leader in renewable and circular solutions.

In 2022, Neste reached EUR 25.7 billion in revenue and employed 5244 employees on average. The company produces renewable products at its refineries in Finland, the Netherlands, and Singapore entirely from renewable raw materials with a current annual nameplate capacity of approximately 3.3 million tons. It plans to expand its total production capacity of renewable products to 5.5 million tons by the beginning of 2024.

5G Innovations for Industry Transformation: Data-Driven Use Cases, First Edition.
Jari Collin, Jarkko Pellikka, and Jyrki T.J. Penttinen.

Published 2024 by John Wiley & Sons, Inc.

Neste is committed to helping its customers to reduce their annual greenhouse gas emissions by at least 20 million tons of CO_2-equivalents by 2030. Reducing emissions and replacing crude oil with renewable and circular solutions is at the heart of Neste's strategy. The company has set an ambitious target to make its Porvoo refinery in Finland the most sustainable refinery in Europe by 2030 and to reach carbon-neutral production by 2035. Neste also aims to reduce the use-phase emission intensity of sold products by 50% by 2040 relative to 2020 levels. These targets will be achieved through Neste's efforts to increase shares of renewable and circular solutions, as well as working with suppliers and partners to reduce emissions across the value chain.

The company is constantly introducing new renewable and recycled raw materials, such as liquefied waste plastic, as refinery raw materials. The renewable raw materials include wastes and residues, such as used cooking oil, animal fat from food-industry waste, vegetable oil processing waste, and fish fat from fish-processing waste. Neste can take in dirtier raw materials than most manufacturers can handle and upcycle them into high-quality fuels.

Our case study focuses on operations at Neste's Porvoo refinery, which is among the most efficient and versatile refineries in Europe, producing more than 100 end products for customers globally. Today, the refinery has a total annual production capacity of approximately 12 million tons, of which some 2 million tons are already produced from 100% renewable raw materials. This is possible thanks to Neste's proprietary NEXBTL™ technology; a unique platform capable of turning diverse renewable fats and oils into premium-quality renewable products, such as fuels and feedstock for producing polymers and chemicals. To meet the company's sustainability targets, Porvoo refinery is in the middle of a transition to becoming a site of fully renewable and circular solutions and ending crude oil refining by the mid-2030s. This is a major transition that will strongly affect the whole production of the refinery throughout the next decade.

11.2 Industry Transformation Challenge

As in many other industries, digitalization is transforming the oil and gas sector. Accordingly, Neste Porvoo refinery has an ongoing digital transformation that is intended to assist efforts to meet its strategic goals (carbon-neutral production by 2035 across its supply chain, optimizing production capacity and efficiency, and improving customer relationships). Digitalization is helping Neste to leverage business value across all these fronts, providing better means to boost sales and marketing, improve sustainability, and increase the efficiency of production and logistics. Digitalization and data are at the core of Neste's strategy. With real-time data, the company can make faster and better decisions, increase efficiency, and stay ahead of the competition while developing new differentiating capabilities for Neste.

Digital transformation is not only about technology: digitalization only produces value when people have learned to think and act in new ways. From a people perspective, this means empowering teams using Agile methods, enhancing their digital and data skills, and promoting close collaboration with customers and partners. Success stories are built together with customers, partners, and employees.

A notable development in Neste's digital transformation of production was the acquisition of a private LTE network (4G) for its Porvoo refinery a few years ago. The network has mainly been utilized for improving basic wireless connectivity in the refinery area. In addition, there have been some testing and use cases to further develop the production processes with the enhanced connectivity. The network is being utilized in efforts (not yet complete) to integrate real-time sensor data from machines, processes, and vehicles in order to realize the full potential of production control systems. Now, industrial 5G technology offers new opportunities and features to automate operations, enable analytics, and improve safety even further.

Complementing current best practices with 4G networks, the case study focuses on new opportunities offered by using drones connected to a private virtual 5G network. The purpose of the study is to investigate the potential uses of 5G-enabled drones in diverse operations and the transition toward more sustainable production. It increases insight into ways to effectively implement 5G-enabled drones in a major industrial sector to improve production performance and safety. It also enhances understanding of ways to effectively capture the value of complementary 5G and drone technologies. Based on interviews with Porvoo refinery personnel, we describe the clearest potential uses of 5G-enabled drones to improve safety, production, and maintenance operations.

11.3 Data-Driven Use Cases

A drone, also referred to as an uncrewed aerial vehicle (UAV), is an aircraft that operates without any human pilot, crew, or passengers on board. Initially developed during the twentieth century, these aerial vehicles were specifically designed for military operations that were deemed too "dull, dirty or dangerous" for humans. By the start of the twenty-first century, they had become essential assets for most militaries. Today, there are several commercially available drone technologies and solutions.

The use of 5G-enabled drones is emerging for diverse industrial processes. They can be operated through private and public 4G/5G networks, utilizing cloud-based services with standard APIs. A typical system consists of drones, a docking station, a ground control station, as well as add-on equipment such as a dual gimbal video camera. It enables the use of multiple drones to fly on automated individual missions steered from a single ground control station running on the mobile edge cloud.

11.3.1 Introduction to the Use Cases

A basic wireless video surveillance system is a cost-effective way to enhance safety and improve efficiency in a large refinery area. 5G-enabled drones with mounted cameras can provide a robust network of aerial sensor footage that greatly enhances situational awareness and decision-making by generating a continuous stream of data on the status and condition of all types of assets and infrastructure. The addition of video analytics creates more valuable applications. 5G-enabled drones can realize many tasks and functions that increase the efficiency and safety of operations. When equipped with cameras they can collect abundant data and video streams from numerous parts of refineries that provide real-time information with high practical value for numerous applications, including maintaining safety, supporting in-field workers, and improving data security, production, and product quality. Figure 11.1 illustrates the technical set-up use cases considered here, showing how real-time data collected from 5G drones can be integrated with a 5G network and mobile edge computing providing an integrated connectivity and computing platform.

Multiple use cases have been identified for 5G-enabled drones to utilize real-time data for improving safety, maintenance, and production processes, all involving the use of the same basic technical components and underlying platform. The technical platform includes drones with attached cameras and other sensors, connectivity, and data storage systems. The differences between the use cases start

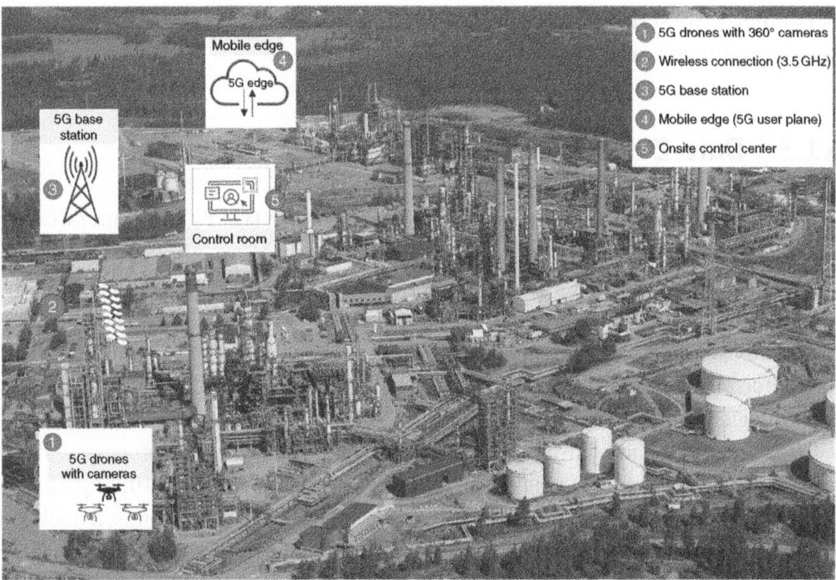

Figure 11.1 The technical setup for 5G drones at the Porvoo refinery. *Source:* Reproduced with permission from Neste.

to become apparent in video analytics (in the mobile edge could use the latest machine learning methods) when the acquired datasets are refined to extract meaningful information. The outcome is sent to onsite mobile applications used by the refinery's control center and technicians in the field.

By analyzing current industry developments and conducting interviews with experts in the industry, we have identified the following potential use cases: leak detection, inspections, asset management, surveillance and emergency response, logistics, and transport. Each of these cases is described in the following sections:

11.3.2 Leak Detection

Leak detection involves various techniques to identify and locate potential issues in tanks, pipelines, and other components of oil and gas systems. Through the combination of infrared cameras, gas detectors, and data analytics, gas escapes, oil leaks, or other emissions can be detected in real-time. Leakages of oil or fugitive gases can not only be harmful to personnel and the environment but also cause losses of raw materials or products. Early leak detection can thus improve a refinery's safety, sustainability, and efficiency. Moreover, reducing methane emissions from oil and gas operations is among the most cost-effective and impactful actions to achieve global climate goals [1].

Figure 11.2 shows an imaginary situation how a gas leak can be detected by utilizing a drone's video camera and video analytics.

Figure 11.2 Imaginary situation on how gas leak can be visualized with thermal imaging.

11.3.3 Inspections

According to statistical data, the global oil and gas market spends as much as US$ 37 billion annually on monitoring and inspections alone [2]. In many cases, production must be stopped for some time to safely inspect them, which increases costs. Drone-based asset management and inspection is not a future technology, but an innovation that is ready to improve operations and reduce these costs. Furthermore, drones reduce needs for personnel to enter hazardous areas without the need to shut down operations. Moreover, drones can limit work at heights and in spaces that would be highly confined, and/or dangerous, for personnel. There are many kinds of inspections that could be potentially improved.

Drones can perform inspections to quickly and regularly acquire visual and 3D LiDAR (Light Detection and Ranging) data on the internal condition of an asset (e.g. pipe racks, storage tanks, vessels, chimneys) to improve planning for maintenance shutdowns and reduce unplanned downtimes. They can detect signs of defects such as coating deterioration, damage, corrosion, or cracks.

Their use in inspections of parts of facilities that are difficult for personnel to access can reduce needs for scaffolding, which is expensive to contract and move around. This also reduces work at heights and in confined spaces (with potentially hazardous gases) for personnel, thereby greatly reducing risks. Additionally, drones enable increases in the speed and frequency of inspections, and some can autonomously fly in enclosed spaces with no signals, then autonomously find and fly out of exits.

To exemplify their utility, their use in flare tip inspections avoids needs to shut down plants, with cost savings in the order of millions of dollars. Flare stacks are hazardous sites for inspection personnel, and some industry players routinely use drones to inspect flare tips and floating roof tanks, as well as to detect abnormalities like steam leaks or loose railings at the top of stacks. They are also valuable for powerline inspections to prevent or trace damage if short-circuits occur, for mapping brushwood and tree branches under power lines to identify optimal times for clearing works, and for monitoring powerline temperatures (which must remain within specific ranges to avoid failures).

11.3.4 Asset Management

The oil and gas sector is asset-heavy and requires robust, cost-effective methods to manage and maintain physical and nonphysical assets to maximize their value and efficiency. Asset management can be assisted by modern technologies that can provide real-time data and insights to optimize assets' performance and maintenance schedules. It involves the prioritization of asset performance, lifecycle costs, and risk management. By managing assets in a professional way, refineries can improve the reliability and availability of their assets, reduce unnecessary downtime and maintenance costs, increase their returns on investment, and ensure compliance with regulations and standards.

5G-enabled drones equipped with high-resolution cameras and sensors can fly over a refinery and capture detailed images and videos of various components, such as pipelines, tanks, and other structures. The visual data acquired can be analyzed in real-time using 5G networks and processing algorithms to identify any defects or damage, which can help maintenance teams quickly and accurately assess the condition of the components and prioritize repairs.

Assets can also be digitally modeled by using 5G-enabled drones. High-definition 2D maps and 3D models of facilities and vertical structures constructed from digitized inspection results can be obtained using survey-grade drones, with numerous benefits. Inter alia, they can increase understanding of maintenance requirements and greatly facilitate the tracking of construction progress, inspection of elements, identification of weak points in structures, detection of deformations, repetition of observation cycles to get reliable information, as well as the creation and sharing of common operational pictures.

11.3.5 Surveillance and Emergency Response

The ability to enhance real-time situational awareness is a major advantage of 5G-enabled drones, as they can be very rapidly deployed to detect changes and anomalies. In addition to previously mentioned uses, they can be utilized to identify unauthorized people or objects entering a facility to prevent sabotage, terrorism, espionage, and other threats to safety [3].

The drones can complement existing video monitoring that provides 24/7 oversight of multiple sites, all from one control room. The use of flexibly moving cameras can allow whole areas to be monitored more effectively and allow faster responses to detected threats, e.g. mobilization of onsite security guards to manage the situation.

They can also monitor and analyze various operational parameters at an oil refinery, such as temperatures, pressures, and flow rates. This real-time information can provide insights into the performance of refinery components, such as heat exchangers and reactors, which can help optimize their operation and increase efficiency while minimizing downtime.

In emergency situations, 5G-enabled drones can be quickly deployed to assess and respond to emergencies, such as fires or explosions, by providing emergency response teams with real-time footage. They can also be used to calculate the extent of damage, identify areas that must be evacuated, and for safety inspections in hazardous areas where human access is limited or unsafe. For instance, a 5G-enabled drone with gas sensors can be flown over areas where there is a risk of gas leaks or spills to detect dangerous levels of gases or other chemicals.

11.3.6 Logistics and Transport

5G-enabled drones can also be utilized in different refinery logistics and transportation-related situations. They can, for instance, assist a port operator in the harbor by creating a basic situation model and rapidly producing extra information in a disturbance situation (e.g. harbor system malfunctioning, difficult weather conditions, high-risk transport, accident, etc.). The drones can also assist in the transportation of small goods or tools through a refinery area.

11.4 Benefits of 5G

Actors in many industries want to use remote-controlled machines such as drones, cranes, and robot arms to boost operational efficiency in hard-to-reach areas and increase safety by keeping workers out of hazardous environments [3]. In particular, energy providers are interested in use cases like drone video surveillance of pipelines, plants, and infrastructure for both safety reasons and business efficiency. Real-time streaming of high-definition video from drones, combined with analytics for detection, can help energy providers locate risks and defects, helping to prevent leaks and other incidents [3]. Figure 11.3 illustrates Nokia's approach how 5G technology integrates flying drones and flight operations at customer's cockpit.

Traditionally, point-to-point Wi-Fi connection is used for remote control of drones and data transmission, with one controller for each drone [4]. However, there are clear limitations in using Wi-Fi technology for moving vehicles when handovers between different base stations are needed. Therefore, Neste tested a private LTE network to fly and operate drones in the refinery area. Although the technology can support several remotely controlled applications in the refinery environment, the need to transmit more data back to the command center is driving demand for higher uplink capacities. This is where 5G technology makes a difference. For example, a private LTE network can support full HD video streams, but a 5G network will be needed to support 4K video and sensors [3]. A 5G connection can also provide higher 3GPP-based data security and reliability [4] and low latency to support increased density of remote applications as well as video analytics for automatic drone control, precise object recognition, or detection of defects.

Machine learning algorithms can identify anomalies in thousands of simultaneous video, audio, and sensor data streams and flag potential incidents in real-time [4]. Teaming 5G video with analytics and AI will enable defects to be detected sooner and more easily. This could be highly valuable at hazardous production sites by allowing product designers and engineers to view real-time video from production lines without traveling to the production site [3].

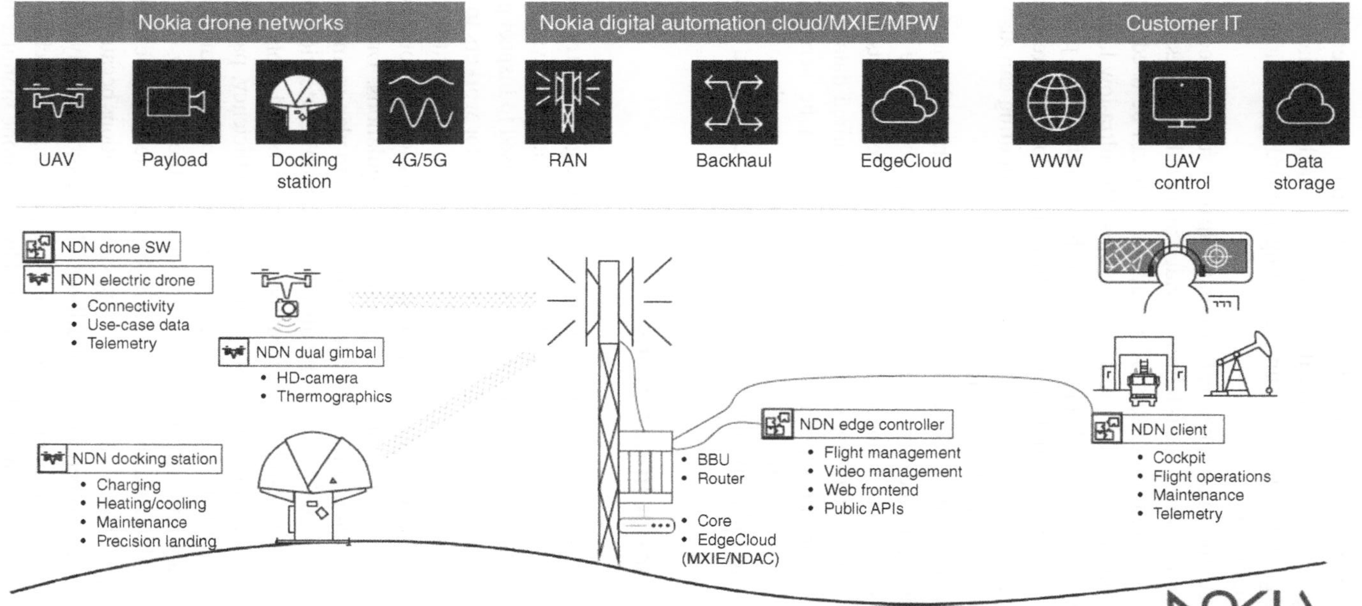

Figure 11.3 Complementary 5G and drone technologies [4]. *Source:* Reprinted with permission from Nokia.

Nokia has developed an end-to-end 5G UAV platform for automated data collection at the edge of the 5G network. It features a certified industrial-grade UAV with integrated graphics processing unit, enabling onboard applications and open APIs for full automation.

Mobile edge computing has key features that help to ensure the necessary quality of service for streaming high-definition video and enable processing of large data volumes generated by sensors. The acquired data can be used not only to monitor assets but also to trigger real-time actions. The ultra-low latency at the edge optimizes mission-critical applications for the manufacturing process. Data processing can shift from central locations in the cloud to local on-premises edges, which reduces enterprises' security concerns. Edge computing can also simplify the integration with other IT systems.

Tangible benefits include reductions in costs, as equipment is monitored (for temperature, pressure, location, humidity, etc.) and problems are flagged preemptively, thus reducing scheduled maintenance and repair costs, as well as unplanned downtime. Refinery staff can also use the acquired data to optimize maintenance and repair schedules. Equipment can then be serviced before problems become serious or irreparable, thereby preventing the occurrence of significant damage and increasing the average lifespan of connected equipment.

11.5 Managerial Implications

5G-enabled drones open huge possibilities for the oil and gas industry to improve refineries' safety and efficiency. They can be used in various ways within a refinery. Drones equipped with cameras and sensors can be used to inspect pipelines, tanks, and towers. This can save time and reduce needs for workers to perform dangerous tasks. They can be used to monitor refineries and identify potential security breaches. They can also be used to monitor emissions and other environmental factors, helping to ensure compliance with regulations. Drones can be used to create 3D maps of a refinery, which can be used to plan maintenance and repairs. The data collected can help identify areas that require attention and prioritize repairs. They can also be used to deliver small parts and tools to workers within the refinery. This can save time and increase efficiency, particularly in large and complex refineries. Overall, the use of drones can significantly enhance the safety, efficiency, and environmental compliance of the refinery.

It should be noted that the use of drones is highly regulated, and the legislation is still being developed. There is a gray area in rules and regulations because the multicopter type of drone has been technologically developing more rapidly than related legislation, but there are major efforts to close the gap. Thus, the anticipated effects of these legislative actions should be considered by both drone manufacturers and their users [5].

References

1 IEA, "International Energy Agency," 2021. [Online]. Available: https://www.iea.org/. [Accessed 20 June 2023].

2 M. G. Dutta, "Energy Giants Can Gain From Investing in Drones. Here's How," yahoo!finance, 8 August 2019. [Online]. Available: https://finance.yahoo.com/news/energy-giants-gain-investing-drones-132701781.

3 Nokia, "5G Use Case eBook," 2020. [Online]. Available: https://www.nokia.com/networks/5g/use-cases/video-surveillance/

4 Nokia, "Nokia Drone Networks Overview," Espoo, 2023.

5 R. Virrankoski, Autonomous Sensing using Satellites, Multicopters, Sensors and Actuators, Espoo: Aalto University, 2023.

Part III

Transforming for Digital Business

12

Industrialization of the Lessons Learned

12.1 Introduction

The lessons learned from the five case studies are summarized in this chapter. We aim to identify and describe contemporary Industrial 5G use cases that drive customer value, productivity, and sustainability in the selected industry ecosystems. Our viewpoint is an industrial enterprise that seeks new business opportunities to leverage 5G technology in its digital transformation journey. The chapter focuses on sharing key lessons and best practices between the industry verticals, as well as understanding which elements are common and which are industry-specific.

Table 12.1 shows the industry case studies and the potential value of 5G technology in their respective industry transformation. Shared lessons and best practices between different industry sectors are an essential part of this analysis.

Based on the lessons learned, we first present the generic composition of an Industrial 5G solution and its main technology components that need to work together seamlessly. This part includes an analysis of which components are industry-specific and which are common to the industry sectors. Next, we describe the main opportunities for industry ecosystems to utilize 5G technology in boosting the use of data-driven industry IoT solutions and related use cases in different sectors. The main opportunities are broken into development areas where 5G technology can enable a major leap in value. Last, we discuss barriers to implementing Industrial 5G solutions. The main barriers are identified and prioritized together with the five industry sectors. Each barrier is described more thoroughly.

Our five case studies, each representing a different industry sector, demonstrate how the convergence of information technology (IT), operational technology (OT), and communication technology (CT) is rapidly changing traditional industrial processes. They become networked, virtual operations that are optimized

5G Innovations for Industry Transformation: Data-Driven Use Cases, First Edition.
Jari Collin, Jarkko Pellikka, and Jyrki T.J. Penttinen.

Published 2024 by John Wiley & Sons, Inc.

Table 12.1 Summary of industry case studies.

Industry	Case study description
Mining	Wireless remote control of autonomous vehicles is already a reality in certain processes – compared to existing wireless Wi-Fi technology, what benefits and new business opportunities does 5G technology enable in building, operating, and maintaining wireless solutions in mine areas – both underground and on the surface?
Forest	How to improve the efficiency of bioproduct mill operations by utilizing 5G technology in production, maintenance, and safety.
Lift	During a building construction project, how to provide flexibly the building ecosystem players with a common, real-time 5G data platform of construction site-related information.
Telecom	How can real-time data from system's components be applied to improve energy efficiency of 5G network, edge computing, and customer industrial processes?
Oil and gas	How can 5G drones and 5G-enabled video analytics improve the reliability and efficiency of sustainable production, maintenance, and safety operations in a large refinery area?

based on real-time data for the benefit of a bigger industry ecosystem. Here, Industrial 5G can provide the ecosystems with a common, trusted connectivity and computing platform. This type of technical integration of cyber-physical systems into the use of industry-wide IoT applications has implications for value creation and business models. It opens up new business opportunities for data-driven Industry IoT applications for the whole ecosystem.

12.2 Industrial 5G Solution with New Opportunities

Industrial 5G refers to an Industrial Internet (sometimes referred to as industry IoT) solution that combines a massive amount of data in real-time from multiple networked sensors using a secure connection and a computing platform, enabling ultra-reliable, low-latency, and high-bandwidth communication in designated locations. Based on 5G technology, it is a secured, wireless data platform dedicated to an industrial enterprise to integrate different industry IoT applications between its ecosystem players.

Any Industrial Internet solution requires an integrated technology infrastructure that works seamlessly together in a secure way [1]. Our case studies demonstrate how operational value is created in distinct parts of the infrastructure. Industrial 5G infrastructure consists of the following generic parts: sensors, connectivity,

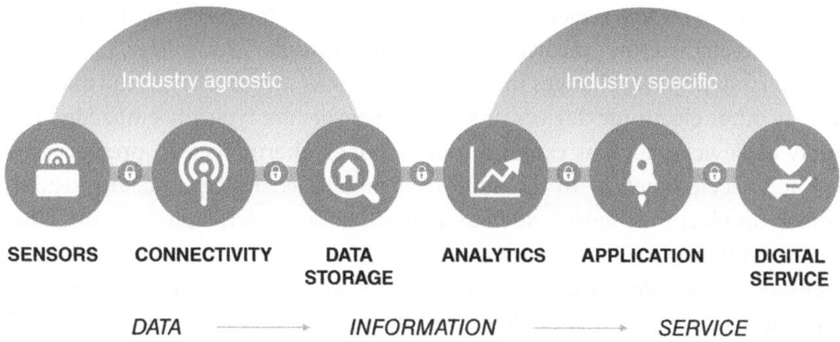

Figure 12.1 Technology Infrastructure for Industrial 5G. *Source:* Collin [2]/International Frequency Sensor Association.

data storage, analytics, applications, and digital services, as well as integrated data security controls [2]. New opportunities to increase operational value lie in each of these parts. The biggest value is created when all these elements, as shown in Figure 12.1, work seamlessly together in a secure way.

In general, technology is no longer a bottleneck, but the challenge is that the parts need to be integrated and work seamlessly together. Our cross-industry analysis shows that sensors, connectivity, and data storage are highly standardized parts. These are common elements regardless of the industry sector – data are collected, transferred, and stored as bits (0/1). This contrasts with analytics, application, and digital services which are elements where data are turned into valuable information that users should access with a well-functioning user interface. These parts are clearly industry-specific – and they include processes where the biggest value is created. A full-scale Industrial 5G solution is, thus, not limited to the connectivity part only but needs to cover other elements too. A brief explanation of all these infrastructure elements, especially from a 5G technology angle, is needed to understand the full potential of opportunities.

12.2.1 Sensors

The first part of an Industrial 5G solution is a sensor, that is, a device used to detect and measure physical phenomena such as light, sound, temperature, pressure, and movement. A sensor is a device that produces an output signal for the purpose of sensing a physical phenomenon, such as temperature, pressure, luminosity, sound, humidity, location, turbidity, or height of the liquid surface [1]. In industrial settings, it typically collects data from machines, vehicles, toolsets, buildings, and assets to which it is installed, or from its surroundings. Sensor data are also used to monitor industrial processes, events, and human behavior.

Sensors are used in a variety of applications, such as industrial manufacturing, automotive engineering, medical devices, and consumer electronics. They work by converting physical changes into electrical signals, which can be read and analyzed by other electronic devices or systems. Sensors are typically embedded in a variety of devices, such as cameras, microphones, thermometers, pressure sensors, and accelerometers. In the context of mobile technology, a device usually refers to an electronic device, such as a smartphone, tablet, laptop, or any other type of computing equipment.

Sensors can be categorized in diverse ways based on the physical phenomenon they detect or the technology they use. They are critical components in many modern devices and systems, as they enable the collection of surrounding data and provide feedback to control systems. Sophisticated sensors are also being developed for emerging technologies such as robotics, artificial intelligence (AI), and autonomous vehicles, enabling these systems to sense and interact with their environment more effectively. Embedded sensor devices often include actuators. An actuator is a type of device that is used to control a physical system by converting an electrical signal into a mechanical action. It is a key component of many diverse types of systems, including robotics, automation, and industrial control systems.

Various sensors and actuators are often embedded into modern wireless systems. 5G technology opens new opportunities to expand sensor coverage and capacity, reduce their life cycle costs, and create new kinds of process monitoring, controlling, and optimization capabilities. For instance, 5G-supported video cameras and drones are being actively assessed to automate industrial processes in large industrial areas. The 3GPP organization plays an essential role in the development and standardization of 5G technology, including sensors and devices.

12.2.2 Connectivity

Connectivity, the second part of an Industrial 5G solution, allows the transfer of sensor data from multiple sources into data storage in real time. Generally speaking, connectivity refers to the ability of devices, networks, and systems to communicate and exchange information with each other. In industrial processes, connectivity is currently realized using various legacy wired and wireless technologies. The recent development of cellular networks and devices (4G/5G) and the introduction of industry IoT applications have clearly shifted the focus to wireless solutions, as they can flexibly integrate connected devices and enable real-time data-driven decision-making with short response times. At present, 5G connectivity only complements the other existing technologies. Although modern wireless solutions are being used in industry applications, there are still many industry connections with wired legacy technologies that comply with specific safety and high-reliability needs.

Certainly, 5G technology will eventually revolutionize this area by bringing ultra-reliable, high-capacity, and low-latency wireless connections that are easily available to a whole industrial area, but it will take time to modernize all connections. 5G connectivity can be arranged either through private or public networks. A private network enables more secured connections for authorized devices and users only and is not publicly accessible. 5G private networks, also referred to as non-public networks (NPNs) by 3GPP standards, are physical or virtual 5G cellular systems deployed for private use [3]. The reliability, capacity, and latency characteristics of a 5G network are critical factors when industrial enterprises are making decisions on their future connectivity solutions.

12.2.3 Data Storage

The third part is data storage, which refers to the process of storing operational data in a physical or virtual location that can be accessed by authorized users. In industrial settings, it requires a secure data center where sensitive sensor data from industrial processes are collected and stored for further processing. There are several methods available for storing data, ranging from a traditional dedicated server for production to public cloud-based storage options. Due to security and business continuity reasons, the solution for OT systems is often based on a physical on-site server managed by the company's local employees. In OT systems, public cloud solutions often do not fulfill all data security requirements nor provide fast enough service levels to handle production disruptions. Although existing production automation systems are gradually being cloudified, the data from these systems are not allowed to move to the internet. A private cloud is a solution for this. In private clouds, the company has complete control over data and their security, privacy, and compliance. It is a virtualized computing environment that provides similar benefits to a public cloud but with additional control, customization, and security inside the company's own firewalls.

However, for basic commercial IT services, a public cloud solution can be used. The most suitable data storage solution, thus, differs for different applications, leading to hybrid solutions. A hybrid cloud is a type of cloud computing environment that combines the benefits of both public and private clouds. It is a combination of clouds bound together by standardized technology that enables data and applications to be shared between them. In a hybrid cloud, sensitive data and workloads may be kept in private clouds, while nonsensitive workloads and applications may reside in public clouds.

Here, 5G technology allows more flexibility to build cloud-based data storage solutions for companies and the industry ecosystems they belong to. A 5G mobile edge is a type of cloud computing that is placed at the edge of a mobile network.

It provides computing resources and services closer to end-users while reducing latency and improving connectivity. With a 5G mobile edge, companies can deploy distributed computing resources and services at the network edge, enabling applications to run locally on the devices rather than in a centralized data center. This reduces the amount of data that needs to be transferred over the network, reducing latency and improving response times, while also increasing network efficiency and reducing bandwidth usage. Additionally, the mobile edge can improve network security and privacy by enabling end-to-end encryption and supporting secure communication protocols.

Overall, 5G mobile edge computing is a promising technology that has the potential to improve the performance, security, and efficiency of many connected industrial applications and use cases.

12.2.4 Analytics

Analytics represents the fourth part of an Industrial 5G solution. It turns real-time data into meaningful information that is essential for fast, fact-based decision-making and forecasting analysis. Analytics refers to the process of analyzing data to gain insights, identify patterns, and make informed decisions based on the findings. It often involves collecting and processing data from various sources, such as data from internal and external information systems as well as a large amount of sensor data from production processes. Descriptive analytics involves summarizing data to gain a better understanding of past events. Predictive analytics, in turn, uses software algorithms to make predictions about future events based on historical data, e.g. in production maintenance and support services. The benefits of predictive maintenance, such as reduced downtime, increased safety, and reduced maintenance costs, are a result of the effective use of real-time data. Here, an Industrial 5G solution plays a central enabling role, as it collects and integrates the different data flows of industrial processes together with short response times.

Analytics is where OT meets IT: the sensor data of automation systems is enriched for the use of management systems. Deep analysis of data patterns with various solutions based on machine learning and AI is utilized for this. In practice, machine learning is an application of AI that enables OT/IT systems to improve their performance automatically by learning from historical data. Machine learning enables the systems to recognize patterns, make predictions, and carry out proactive tasks that improve the performance of industrial processes. For instance, machine learning is used in autonomous vehicles to identify objects, detect collisions, and optimize routes. 5G technology as an integrated connectivity and data platform can play a central role at this crossroads and enrich analytics with communications network data.

12.2.5 Application

The fifth part of the solution is a fit-for-purpose application that utilizes the enriched information from analytics and provides users with handy mobile apps with a well-suited user interface. Usually, these kinds of industry IoT applications are highly industry-specific, as they are designed to serve specific needs and requirements of the industry. The users of these applications – including people, machines, and vehicles – are often designed to carry out specific process tasks that utilize and complement the information of other industrial processes. The handling of real-time process data becomes a critical factor. Here, 5G technology can make a big impact, as it provides these kinds of applications with reliable low-latency wireless connections.

These industry IoT applications should be aligned with the company's enterprise architecture and API strategy in order to integrate them into the digital services of an industry ecosystem. Already today, some industry sectors such as the mining and forest industries utilize modern digital services based on virtual environments. Virtual reality (VR) and augmented reality (AR) applications can drive automation and intelligence into industrial processes and operations at a massive scale.

AR enhances the user's perception of the physical, real world with a virtual environment. AR applications typically utilize data from OT/IT systems, sensors, cameras, and GPS to detect the user's location and movements and to align digital content with the real-world environment. Nowadays, these applications are being used by maintenance workers in the field and by machine operators in control centers. VR applications create a digital interactive environment that simulates the real-world environment or creates a completely imaginary world. They can provide users with a sensory experience that feels and looks real. 3D modeling and digital twins are existing use cases where VR applications are being actively utilized. Again, utilizing these kinds of virtual applications in mobile use requires real-time capabilities that 5G technology enables in terminals and networks.

12.2.6 Digital Service

Digital service, the final part of an Industrial 5G solution, acts as a productized and sellable service for the industry ecosystem, providing users with digital content, tools, or resources. It may include web-based or mobile applications, online resources, online shops, subscription-based services, and other digital offerings. This part ties together the whole Industrial 5G solution, allowing new digital business models based on data-driven industry IoT applications. These digital services complement the existing product-based business models by integrating and leveraging the power of digital technologies across the industry ecosystems. They also provide industrial enterprises with new growth opportunities to expand their offerings and reach new markets.

Our case studies demonstrate that opportunities lie in all elements of the 5G infrastructure – not only in the connectivity part. Sharing industrial process data among industry ecosystem players adds value. Customer value is increased when data are turned into meaningful information and digital services for the benefit of the whole industry ecosystem. The greatest value lies in the development of sellable digital services. Industrial 5G as an integrated connectivity and computing platform opens new opportunities to create and operate for industry IoT applications.

12.2.7 Integrated Solution with Data Security

All infrastructure parts of an Industrial 5G solution must seamlessly and securely work together, as seen in the case studies. Using a single part without integration does not really pay off. For instance, upgrading only the connectivity part to 5G will not yield the full benefits, so the whole value creation process from turning real-time data into meaningful information and valuable services for the respective ecosystem should be considered. It is, therefore, important to link technology upgrades to the company's own digital transformation. However, to begin, it is wise to have a "start small, scale fast" deployment approach and first create an end-to-end demonstration with a single use case with limited scope, e.g. in a mine, a mill, or a construction area.

Although each industry sector has its own unique characteristics, many commonalities and synergies across industries do exist. Sensor, connectivity, and data storage are parts where common, standard technologies are easily available and synergies are justifiable. These parts could be procured as standard products and services – or as an integrated turnkey solution. In turn, data analytics, applications, and digital services are clearly industry-specific parts where customization is often needed. However, these parts also generate the biggest customer value. For these components, there is no standard, common solution available in the marketplace due to significant issues with the integration of differently scaled systems.

Finally, security is an important aspect of the 5G technology, covering networks, devices, and applications. Industrial 5G solutions must cope with data-related risks: data security, data privacy, data compliancy, and cyber threats. 5G has inbuilt features to tackle these risks, but still, these risks have to be continuously taken into account while building the solution and running the operations after the deployment. Network architecture involves creating secure zones, controlling access, and monitoring network traffic to identify and isolate suspicious activities. Network slicing is a technique to partition a physical 5G network into multiple virtual networks improving the security of the network. The use of encryption to protect data in transit and at rest is critical to ensure the security of 5G networks. 5G devices need to have security features that are built-in and updated on a regular basis. This includes strong authentication mechanisms, encryption, and security protocols.

Based on the lessons learned, we next elaborate on three alternative approaches to building and operating Industrial 5G solutions from a network viewpoint: private network, virtual private network, and network slice, and discuss selected major opportunities that 5G technology brings to the industrial processes.

12.3 Alternative Approaches to Industrial 5G

This section summarizes the alternative approaches to building and operating Industrial 5G solutions utilizing the new capabilities of 5G technology. Industrial enterprises do not always need to invest in their own 5G network to obtain Industrial 5G solutions. Optionally, they can utilize public 5G networks, depending on the requirements and needs of the solution. The alternatives are summarized in Table 12.2.

Each alternative is thoroughly explained next.

12.3.1 Private Networks

A 5G private network is a wireless network built on 5G technology and dedicated to a specific industrial enterprise in a selected location. Access to the network is restricted to predefined users inside company firewall settings without unwanted external access. The network infrastructure is owned, managed, and operated by

Table 12.2 Alternative approaches to Industrial 5G.

#	Alternatives to Industrial 5G	Description
1	Private network	An industrial enterprise considers full autonomy as mandatory and invests in its own private 5G standalone network on the industry site to ensure maximal business continuity, data security, the lowest latency possible, and network flexibility.
2	Virtual private network	An industrial enterprise considers data security and low latency such crucial factors that they invest in a 5G mobile edge in proximity to the industry site (public network).
3	Network slice	An industrial enterprise considers network capacity such a crucial factor that they pay a premium for the improved 5G service levels on the industry site (public network).

the enterprise itself. Private networks allow industrial enterprises to have greater control over their network resources and security, enabling them to customize their network to meet their specific requirements. 5G technology provides the network owner with improved security features such as encryption, network segmentation, and enhanced user authentication.

Industrial enterprises that invest in their own private 5G standalone network consider full autonomy as mandatory. They want maximum business continuity, data security, the lowest latency possible, and network flexibility on the respective industry site. This enables industrial applications that need high bandwidth and low latency to operate seamlessly. The private network allows customization of the network architecture and the addition of new services as needed.

The performance benefits of the private network guarantee reliable, fast connectivity for users and devices, ensuring that critical data can be transmitted quickly and securely. Private 5G networks can provide users with lower latency than public 5G networks, as the computing of user data takes place at the site location. This is important for industry IoT applications that require real-time responses, such as autonomous vehicles and process automation.

Our industry case studies confirm the growing interest in 5G private networks among industrial enterprises. These networks are likely to become increasingly common over the coming years. Currently, in Germany alone, there are more than 320 private networks utilizing 5G technology [4].

12.3.2 Virtual Private Networks

Another, lighter approach to utilize the benefits of private networks is a virtual 5G private network, that is, regionally isolated network capacity for a specific industrial enterprise. This approach utilizes a 5G public network and its mobile edge computing capabilities in proximity to the industry site. By doing this, network capacity with relatively low latency can be guaranteed for the industrial enterprise. Another important aspect is that the connections take place between the industry site and the mobile edge, ensuring the data are not transferred through the internet. Most data security requirements can be managed with this arrangement.

A virtual 5G private network is a secure and encrypted connection that allows users to access a private network over a public network. It creates a private network by using encryption and tunneling protocols to create a secure connection between two points, allowing users to connect to a remote network securely and anonymously. A dedicated network capacity is reserved for enterprises without a major investment in their own local private network on the industry site. The latency can be kept much lower than on normal public networks since the computing takes place in a mobile edge that is located in close proximity to the site.

Overall, virtual 5G private networks are a secure and more cost-efficient way of having dedicated connections than a company investing in its own network. This approach is suitable for industrial enterprises that consider data security and low latency crucial factors but are able to utilize the mobile edge capabilities of a 5G public network.

12.3.3 Network Slice

The most cost-efficient and simplest approach to obtain limited 5G private network capabilities is network slicing, a new feature in 5G technology. It allows a dedicated network slice, or capacity, to be reserved for a specific enterprise at a certain location and/or for a time period. This is an efficient and flexible way to reserve capacity on a public 5G network. Network slicing is a network architecture that enables the multiplexing of virtualized and independent logical networks on the same physical network infrastructure [5]. Each network slice is an isolated end-to-end network tailored to fulfill diverse requirements requested by a particular application. The network slicing does not guarantee specific latency and safety requirements, as the service levels are defined based on the public 5G network.

Network slicing as a technique can create separate and individual virtual networks within a single physical network. It allows the network to be partitioned into multiple logical networks with its own set of resources and capabilities that can be tailored to meet the specific requirements of different applications and services. Each slice has its own unique characteristics, such as bandwidth, latency, and security levels that can be customized according to the needs of a particular service or application.

Network slicing is a flexible and efficient approach to support a wide range of use cases. It allows communication service providers to allocate resources dynamically to meet the requirements of different services, applications, and users in real time as demand changes. For example, a network slice can dynamically be created for an industrial enterprise that needs a high-performance and dedicated network for business-critical applications at its specific industry site location or locations. This new network capability enables improved quality of service with guaranteed network performance levels.

Overall, network slicing provides a flexible and scalable way to build and use new capabilities in public 5G networks. This approach suits industrial enterprises that consider network capacity as a crucial factor in fulfilling their wireless networking needs. Improved 5G service levels to specific industry sites, however, have an additional cost that needs to be taken into account when comparing the different approaches.

12.4 Barriers to Implementing Industrial 5G

The implementation of Industrial 5G also involves barriers and is anything but a straightforward initiative, as it touches many industrial processes and system areas [2]. It enforces the integration of processes and systems between OT and IT as well. Traditionally, industries have seen and handled OT and IT as two different specific domains, keeping separate technology stacks, protocols, standards, management, and organizational units [6]. The modernization of legacy processes and tools may lead to overwhelming complexity, delays in implementation, and cost increases. Cybersecurity risks and insufficient competencies can also become a bottleneck for the implementation.

In each industry case study, implementation barriers were identified and elaborated on in discussions with stakeholders at the management and operational levels. The main barriers identified in each industry sector are summarized in Table 12.3.

We used these findings in our common workshop and prioritized them together to learn and share best practices between the industry sectors. The prioritization was carried out in a simple manner by utilizing the known *100-Point Method* that is globally used in agile software development to prioritize product backlogs in a scrum team. Basically, the method is a voting scheme of the type that is used in brainstorming exercises [7]. A summary of the common prioritization is shown in Figure 12.2.

The biggest barrier is Competencies. The second biggest one is Change Complexity. The third is Technology, which is followed by Total Cost of Ownership (TCO) in fourth place. Surprisingly, the smallest barrier is Data Security and Privacy. After voting, each barrier and its ranked position was briefly further elaborated.

Table 12.3 Main barriers to Industrial 5G.

	Industry sectors				
Identified barriers to Industrial 5G	**1**	**2**	**3**	**4**	**5**
Technology	x		x		x
Total cost of ownership	x	x			
Competencies	x	x	x	x	x
Change complexity		x	x		x
Data security and privacy	x	x		x	x

Source: Collin [2]/International Frequency Sensor Association.

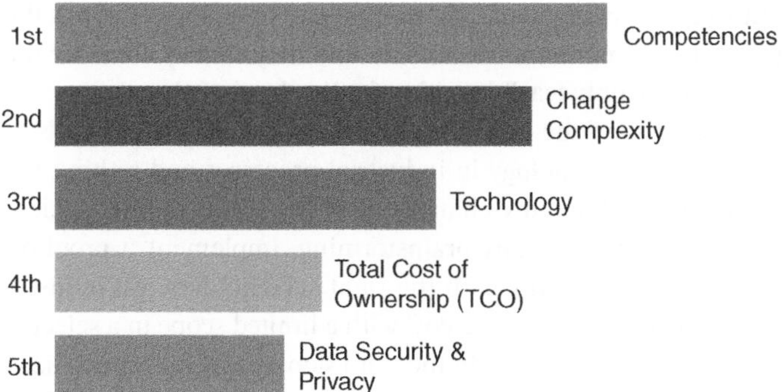

Figure 12.2 Prioritization of the main barriers to Industrial 5G. *Source:* Collin [2]/ International Frequency Sensor Association.

12.4.1 Competencies

All industry sectors consider the main barrier to be a lack of competent people who can design and build use cases to maximize the business value of Industrial 5G. This includes an understanding of the benefits and potential use cases of 5G in industrial applications, as well as the technical expertise to implement and maintain the necessary infrastructure. The biggest challenge is how to merge business and technology perspectives together to discover operational innovations that turn real-time data into meaningful information and valuable digital services. The competence gap is related to the implementation and use of 5G technology in the respective industrial sector. The gap can limit the adoption and benefits of 5G technology in industrial settings.

Closing the competence gap requires training initiatives focused on Industrial 5G technology and its applications. This could include specialized training programs for production managers and IT/OT professionals as a part of digital transformation change programs. Close collaboration between industry leaders, policymakers, and universities to develop and share knowledge can also help to address regulatory and infrastructure challenges that may limit the adoption of Industrial 5G.

Overall, closing the competence gap is essential for the unlocking of the full potential of Industrial 5G and ensuring that enterprises are equipped with the necessary skills and knowledge to succeed in industry digital transformation.

12.4.2 Change Complexity

The implementation of an Industrial 5G solution is seen as a major transformative initiative impacting company's processes, systems, and data models. Therefore, the implementation should be linked to the company's strategy, enterprise architecture, and digitalization journey. Managing change can be complex, especially when there are multiple factors involved. The complexity of changes defines the correct change approach. All representatives of the industry sectors consider the evolutionary approach as the only feasible option for this kind of change.

The key to success consists of the following four steps [1]. The first step is to identify opportunities for 5G technology in industrial processes and rethink the operations strategy. Before making any changes, it is important to have a clear vision of what to achieve. Second, start brainstorming, implement a proof-of-concept (PoC), and analyze the results with the right stakeholders. All industry sectors prefer to start with a pilot to test the PoC with a limited scope in a selected industry site focusing on a single use case. In the third step, strategy is reprioritized based on the lessons learned from the pilot. The solution can be "productized" for a wider implementation after the pilot and start leading changes as a transformation program. Finally, it is important to share best practices continuously and avoid pitfalls during the transformation journey.

12.4.3 Technology

Three industry sectors consider technology as a potential barrier to implementation, especially in industry sites equipped with legacy technology. 5G as a technology is standardized, but industry IoT technology is not. Therefore, an all-encompassing technical package is not available, but system integration is needed. There are many different IoT technology stacks in the marketplace, and enterprises need to integrate the different technical components together. This makes the implementation even more labor-intensive. Furthermore, combining different worldviews of IT and OT easily becomes a barrier.

All industry sectors, however, consider that technology must not become a showstopper for implementation, as a constant renewal of technology is essential.

12.4.4 Total Cost of Ownership

TCO is a financial estimate that calculates the overall expenses associated with owning a particular asset over its lifespan. This includes not only the purchase price of the asset but also the cost of maintenance, repairs, and any necessary upgrades or replacements.

The total costs of an Industrial 5G solution, covering the whole life cycle of the system, is seen as a potential barrier to large-scale implementation by two industry sectors. The value of having a dedicated connectivity and computing platform differs significantly for each industry sector. Therefore, it greatly depends on the industry ecosystem as to whether the TCO pays off.

In the evolutionary implementation approach, the cost is not considered to be a significant challenge for any sector. There are different approaches to Industrial 5G, among which TCOs differ greatly. At this point, cost levels are not considered a major showstopper. On the contrary, Industrial 5G is expected to improve overall process efficiency.

12.4.5 Data Security and Privacy

Many organizations today face significant challenges in maintaining data security and privacy, particularly as new technologies such as cloud computing, mobile devices, and IoT create new vulnerabilities. Therefore, data security and privacy are two critical aspects of Industrial 5G. Data security refers to the measures taken to protect data from both internal and external threats. Data privacy, on the other hand, refers to the rights of individuals to control how their personal information is collected, used, and shared.

Data security and privacy are identified as major risk areas in all industry sectors. The risks are not seen as dependent on Industrial 5G alone, as they need to be managed anyway. In fact, 5G technology can mitigate the risk level, as enterprises can build solutions based on private networks and edge computing.

12.5 Conclusions

The main value of Industrial 5G is created when data are turned into meaningful information and digital services in real time. Improvement opportunities from 5G technology lie in all infrastructure parts, although the core new capabilities are focused on connectivity. The generic parts that are common across industries are sensors, connectivity, and data storage. The components are fairly standardized already. However, the biggest customer value is generated from data analytics, applications, and digital services that are highly industry-specific and customized. For these parts, there is no standard, common solution available. Therefore, the implementation may often require extra system integration work.

The implementation of Industrial 5G involves obstacles and challenges that need to be proactively managed. The main barriers, in order of priority, are competencies, change complexity, technology, TCO, and data security and privacy. Managing these barriers proactively is essential for a successful implementation.

References

1 J. Collin and A. Saarelainen, Teollinen Internet, Helsinki: Talentum Pro, 2016.

2 J. Collin, J. Pellikka, and J. T. J. Penttinen, "Industrial 5G to boost data-driven IIoT applications – Opportunities and barriers," in *3rd IFSA Winter Conference on Automation, Robotics & Communications for Industry 4.0/5.0 (ARCI' 2023)*, Chamonix, 22–24 February 2023.

3 M. Attaran, "The impact of 5G on the evolution of intelligent automation and industry digitization," *Journal of Ambient Intelligence and Humanized Computing*, vol. 14, pp. 1–17, 2021.

4 Bundesnetzagentur, "Übersicht der Zuteilungsinhaber für Frequenzzuteilungen für lokale Frequenznutzungen im Frequenzbereich 3.700-3.800 MHz," 17 June 2023. [Online]. Available: https://www.bundesnetzagentur.de/SharedDocs/Downloads/DE/Sachgebiete/Telekommunikation/Unternehmen_Institutionen/Frequenzen/OffentlicheNetze/LokaleNetze/Zuteilungsinhaber3,7GHz.pdf?__blob=publicationFile&v=9. [Accessed 17 June 2023].

5 A. J. Zharabad, S. Yousefi and T. Kunz, "Network slicing in virtualized 5G Core with VNF sharing," *Journal of Network and Computer Applications*, vol. 215, p. 103631, 2023.

6 C. Giannelli and M. Picone, "Industrial IoT as IT and OT convergence: Challenges and opportunities," *IoT*, vol. 3, no. 1, pp. 259–261, 2022.

7 D. Leffingwell and D. Widrig, Managing Software Requirements, Addison Wesley, 1999.

13

5G Private Network Guidelines for Industry Verticals

13.1 Introduction

As discussed in Chapter 4, there are various private network deployment models. The entities offering private network services to their end-users need to understand the implications of each deployment model in terms of technicalities, ownership, and respective benefits and obligations, as well as techno-economic impacts.

Whether the entity is an enterprise, factory, or other type of vertical representative, the first task in their assessment of feasible deployment models is to understand how well different models can comply with the requirements, e.g. the level of privacy, own control, and latency, among others. These key requirements set the high-level priority for the models. The more demanding and numerous the requirements, the more likely the requirements will not result in a completely ideal deployment model because trade-offs between requirements are often necessary.

Once the deployment models are shortlisted in the order of compliance with the requirements, disregarding the ones that do not meet the requirements well enough, the suitability of the remaining models can be assessed for techno-economic feasibility in terms of the initial and longer-term costs to set up and operate a private network.

13.2 Evaluation of the Requirements

The feasibility of each non-public network (NPN) deployment depends on the foreseen use cases. As an example, the roaming requirements impact the needed coverage and mobility of the end-users of the NPN. This is one of the examples that can dictate the selection between standalone (SA) and PLMN-assisted scenarios.

5G Innovations for Industry Transformation: Data-Driven Use Cases, First Edition.
Jari Collin, Jarkko Pellikka, and Jyrki T.J. Penttinen.

Published 2024 by John Wiley & Sons, Inc.

13.2.1 SNSP Deployment

The main aspects of the previously presented deployment models for standalone NPN (SNPN) scenarios are summarized in the following list, with statements on their suitability for selected use cases.

5G SNPN: OT operates the NPN and its services behind a firewall independently. Provides good security isolation, e.g. to IIoT applications as data is not exposed externally. Optional PLMN interconnectivity through a firewall. The operation and management of SNPN requires a sufficient skillset from the OT company. Provides the opportunity to build a very secure environment but can be more expensive than a partially owned or completely outsourced network.

5G SNPN with SLA: The agreed SLA level impacts the business case: the higher the SLA, the more costly the CAPEX and OPEX due to, e.g. active-active network mode and reliability of, e.g. 99.999% as per 3GPP Rel 15 URLLC performance, or up to 99.9999% as per Release 16 performance, for e.g. TSN interconnectivity to serve critical IIoT applications.

PLMN with Local Infrastructure: This is a special case of PLMN that isolates part of the infrastructure and spectrum for the sole use of private network devices in a limited geographical area. It can be set up technically, e.g. by barring the access from everything other than defined subscribers.

PLMN on Part of an MNO's Licensed Spectrum: Spectrum is typically a very expensive investment for the license holder, so dedicating part of it needs to be designed carefully applying techno-economic optimization, in order to balance the cost and expected quality adequately.

SA NPN on unlicensed spectrum: The radio deployment and RAN business case are both easy to implement as there is no license fee involved. Nevertheless, the QoS cannot be designed due to the load on the shared spectrum from other possible users.

13.2.2 PNI-NPN Deployment

The following list summarizes key aspects.

- *NPN-shared RAN*: NPN and PLMN share part of the RAN, but NPN communications stay within the defined premises. 3GPP defines the technical RAN-sharing options that can be applied well in this model.
- *NPN-shared RAN and CP*: NPN and PLMN share the RAN for the defined premises, while PLMN has network control; the NPN traffic remains within the defined premises. Network slicing serves this model as per 3GPP specifications, complemented by industry form guidelines for slice template setup. Alternatively, the setup can be based on access point name (APN).

- *NPN hosted by public network*: PLMN and NPN traffic are external to the business area so that these traffic flows are served by different networks, and the NPN subscribers are effectively public network subscribers.

13.2.3 Pros and Cons of Deployment Models

Table 13.1 summarizes the key pros and cons of each presented private network variant.

Table 13.1 Comparison of selected private network options.

Private network model	Pros	Cons
SNPN	Access for customization and independent control; high security through full isolation; RAN functions are within reduced geographical area, favoring low-latency applications.	Deployment cost; expertise required for deployment; dedicated network for sole enterprise includes the cost of the whole system in the geographic area.
Shared RAN	Optimizes RAN costs; internal data remains in NPN for good protection; data can be delivered to PLMN as per need. Within the NPN, part of the base stations can be connected to NPN-PLMN comms, while the rest remains internal. Licensed spectrum minimizes interference; deployment is less expensive than SNPN; local functions favor low-latency applications.	External interference can be higher than in SNPN; the overall control of the network is less independent; need for local expertise, although less than in SNSP.
Shared RAN and CP	Licensed spectrum for controlling interference; lower deployment expenses compared to SNPN and PMI-NPN; SLA can be applied between the NPN and public network.	Less independent from public networks; latency typically higher than in SNSP and PNI-NPN deployments; some local expertise required.
Hosted solution (network slicing by NSP/MNO)	Facilitated by NSP, no need for local expertise; fast to set up and adjust based on expressed requirements.	Less control for adjustments as the NS is controlled by the NSP; technology is not yet final; practical deployments require SA 5G network, of which there are currently few.

(*Continued*)

Table 13.1 (Continued)

Private network model	Pros	Cons
Open RAN as an NPN service	Cost can be low; easy to set up by provider; can be hosted as "light-weight" 5G SNPN or NPI-NSN.	Technology is still evolving and products for Open RAN-based NPN may take time to be realized.
5G FWA	Customer Premises Equipment (CPE) is easy to install; replaces fixed cabling.	Use case is adequate in a home office environment but limited for larger enterprise NPN use.
Non-5G-based solutions	Wi-Fi hotspot deployment is easy within the enterprise area; coverage does not require a license, and it can also be extended to reach any Wi-Fi device external to the enterprise premises.	Low security; limited mobility; lack of QoS.
LPWAN	Options based on cellular and non-cellular radio and integrated cellular-based LPWANs are easy to deploy.	Non-3GPP LPWANs have varying security and protection levels, and they require a separate infrastructure.

13.3 Techno-economic Optimization Modeling Aspects

13.3.1 Current Deployment Models

Each NPN deployment model has its advantages and disadvantages. By understanding the requirements of the use cases and applying techno-economic optimization assessment that considers the key variables, the selection of the most acceptable deployment model will ensure a favorable outcome.

In the selection of the deployment model, the task is to understand the realistic needs of the final users of the NPN access, performance, security aspects, mobility, capacity, QoS, and other key factors, and how they can be served by applying cost-efficient technical solutions. It is also important to understand the changing requirements in the predicted future because, along with the evolution of the environment originally selected, an initially optimal model can turn out to be less optimal in the long term.

As described in [1] related to NS scenarios, the collection of the verticals' requirements can be achieved using practical means to interpret the needs of the end-users. From the operators' perspective, this can be a somewhat challenging task, as the verticals may often express their requirements using non-standard terminology, or not be able to formulate the actual requirements. The methodology in [1], despite its original focus on NS, can be useful to be extended to cover additional aspects in other NPN deployment models. This methodology is useful for interpreting the practical vertical needs based on the predicted use cases and for forming technical requirements that can finally be mapped to represent input parameters for the NPN modeling, whether for a tailored 5G NS template for which the GSMA PRD NG.116 serves as a base, a more "traditional" SNPN setup, or a shared network model. The aim of the requirement list is to ensure a common understanding of the environment and also set the expectations for service levels.

13.3.2 On Further Techno-economic Optimization

The assumption of the modeling is that the NPN is a business contract between an entity capable of deploying adequate wireless networks (such as MNO, NSP, network equipment vendor, or system integrator) and an enterprise wanting to facilitate mobile communications for their end-users in their communications through Industry IoT applications (such as port or energy company monitoring and controlling their workflow using intelligent sensor networks).

The model for the selection of the most adequate NPN deployment option can build upon these modular elements:

- *Interpretation* of the enterprise and end-user needs (e.g. using a survey) to form technical requirements statement as a basis for the input parameters, e.g. capacity (number of expected users/devices), coverage, QoS, need for roaming/local-only utilization, services (IoT, voice, etc.).
- *Business aspects* (assessment of the financial potential for investment in terms of CAPEX and OPEX, flexibility for initial and longer-term investments).
- *Forecast*: current, near-future, and longer-term outlook for the possible need for expansion of the network, capacity, and evolving QoS (which is important to avoid investing in multiple types of NPN as the requirements evolve).
- Any other relevant information on the deployment aspects.

The assessment of the feasibility of the deployment options relevant to the scenario under evaluation can be carried out based on these results in a comparative manner. The basis for the economic assessment is the cost for enterprise in terms of the CAPEX (initial deployment and forecasted subsequent need for new infrastructure investments) and OPEX (yearly cost in order to operate NPN).

The cost estimate of each scenario presented in the NPN Deployment Models Section can consider key attributes such as area of deployment, device number (total expense), and radio performance indicators, which together result in the required bandwidth and number of radio cells, and finally in the total cost of cells. As an example, in a very high data rate scenario requiring a large indoor and outdoor NPN, the number of mm-Wave small cells, each resulting in e.g. 80–100 m cell range, can be in the range of dozens per km^2. Other attributes consider the cost of transmission network and spectrum.

For the core network, the cost includes the licenses and other cost items to activate the needed network functions (NFs). The cost of the needed applications/services refers to the support of, e.g. voice service (that requires either its own or outsourced IMS core for integrated voice over new radio), the IoT service license, and location-based services (LBS) deployment. Roaming and interconnectivity cost is related to the agreements with national and international networks. The cost of other variable items can include, e.g. the actual installation of radio access network (RAN), transport network (TN), and core network (CN) equipment, including antenna systems, base station shields, cabling, and any other expense that is incurred while setting up the NPN for enterprise.

For the estimate of yearly operating costs, the same main components as presented in the CAPEX analysis generate expenses such as licensing fees and electricity consumption, whereas an additional item to be considered in operations is the maintenance cost.

It should be noted that the variables can have non-linear interdependencies. As an example, the volume discount of the number of radio cells can also lower the relative cost of core software and feature licensing.

The assessment produces a statement of the suitability of each deployment scenario in terms of how well they can cope with the interpreted requirements. This method can be visualized in terms of the total cost per area as a function of time, considering attributes of interest such as maximum supported device number or maximum data rate. The method serves thus to estimate the initial and longer-term cost of each deployment model under evaluation; it is possible that the initially most cost-efficient option might turn out to be less optimal in the longer run.

13.4 Enterprise/Vertical Requirement Interpretation

As indicated in the previous section, each private network deployment model has its advantages and disadvantages in terms of the techno-economic aspects. In order to set up the most desirable variant, for the one deploying the network, it is important to understand the technical needs of the verticals and the use cases that the selected private network serves. By assessing the requirements and by applying tecno-economic optimization, considering the key attributes and their desired value ranges, the selection of an adequate deployment model is possible, which is favorable to the business.

The aim of the requirement list of relevant attributes is to ensure a common understanding of the environment amongst the stakeholders setting up and operating the private network and possibly sharing parts of the infrastructure. It is beneficial to set the expectations for service level amongst the stakeholders (customers and communication/service providers).

An example of the private network variant is an MNO-hosted NPN using a dedicated network slice (NS). In order to design the NS, the NS provider can apply the principles detailed in the GSMA PRD NG.116. [2] In these types of scenarios, the customer needs are formed as a network slice by using:

- *Generic Network Slice Template (GST)*: set of attributes that can characterize a type of NS/service. The GST is a generic representation of the NS, and it is not tied to specific network deployment.
- *Network Slice Type (NEST)*: the GST filled with values.

Figure 13.1 depicts the principle for forming a GST and NEST.

As indicated in [3], the GSMA Network Slicing Taskforce (NETSLIC) researched regional vertical needs related to use cases of interest and designed a Network Slicing assessment process that has the following steps:

1) *Priority Verticals*: this step is fundamental to identify verticals interested in network slicing and their significance in the markets. The landscape and requirements of these Priority Verticals represent regional needs and can serve as a very useful basis for forming NSs that cover a large set of other similar verticals.
2) *Vertical Needs Assessment*: collection of the needs of the Priority Verticals to handle the communications for their use cases.

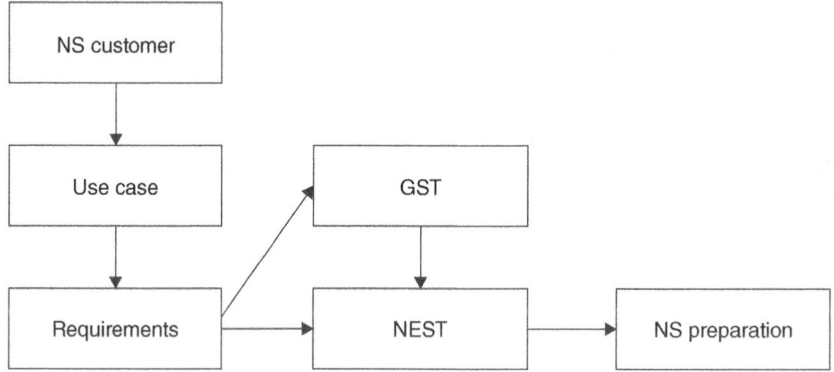

Figure 13.1 The process for forming a network slice instance as per the GSMA PRD NG.116.

3) *Use Case Assessment*: through the reach-out to and assessment of verticals, the practical needs and challenges of their communications environment become more understood.
4) *Network Slice Template Forming*: the needs of the verticals can now be formed to map the relevant attributes and their values into network slicing templates (GST and NEST).

Although the aforementioned method is designed to form relevant NSs and to provide network performance figures for commonly needed use cases, the methodology can also be extended to cover important aspects in the assessment of the most feasible private network deployment models in terms of the interpretation of the needs that can be converted to technical requirements and priorities.

13.5 Enterprise and Vertical Requirements Assessment

The most suitable private networks comply with the high-level requirements of enterprise – or other interested entities using private networks – as closely as possible, such as the level of privacy and freedom to modify the settings of the private network provider.

Table 13.2 presents an example of the items an entity might want to consider and a statement of the relevancy of selected attributes. This table considers the variants relevant to 5G, but it can be extended to cover other mobile generations and technologies. It should be noted, though, that the compliancy statement as presented in this table can vary greatly, depending on the more detailed deployment assumptions. Some of the key aspects impacting these include the spectrum (unlicensed versus licensed), and achievable SLA. It also should be noted that

Table 13.2 An example of the requirements assessment criteria of private networks.

Type of 5G PN	Complexity	Achievable privacy	Achievable reliability	Suitability for *low*-latency comms	Requirement for in-house expertise
SNPN	High	High	High	High	High
PNI-NPN, shared RAN	High/mid	Mid	Mid	Mid	High/mid
PNI-NPN, shared RAN + CP	Mid	Low	Mid	High/mid	High/mid
NPN, PLMN-hosted (NS)	Low	Mid	High/mid	High/mid	Low

Table 13.2 presents an example of a set of parameters, but the values may vary and there may be more parameters depending on the vertical and use case.

Table 13.2 shows which of the 5G NPN variants might be the most suitable depending on an enterprise's desire to achieve certain performance figures. An example of this is a requirement for the lowest possible latency, which could be achievable by deploying a completely isolated and local SNPN (because the infrastructure and services are deployed near the users within the private network boundaries). Nevertheless, if the entity does not have, nor desires to invest in, in-house expertise, this model might not be the most optimal, since in a completely hosted or shared RAN model, the requirement for in-house expertise is lower. Another example is the enterprise's requirement for freedom of customizability (or the time it may take to complete customization in the network setup). A network model managed by enterprise completely in-house (SNPN) complies best with such a requirement.

Once the entity interested in managing a private network has prioritized or narrowed down the variants that best comply with the selected requirements, a techno-economic assessment of these shortlisted variants can be carried out. This assessment considers CAPEX for the initial deployment and estimated need for new infrastructure investments, and OPEX for foreseen yearly costs of operating an NPN. The outcome is the initial and yearly cost of the network.

In the nominal assessment, the CAPEX and OPEX parameters can relate merely to the most relevant and significant items, whereas, at a more detailed level, the model can also consider interdependencies of the items resulting from, e.g. volume and long-term commitment discounts, as well as potential return on investment (RoI), such as in the scenarios where NPN also serves external inbound roamers for additional revenue streams.

It is important to assess the scenarios over a longer period of time because the initial investment as well as the RoI may be very different compared to a mature private network. The outcome of this assessment indicates the costs and RoI figures per stakeholder of interest (Figure 13.2).

13.6 Business Model for 5G Private Network

The model proposed in [4] considers the following parameters:

- Device number, total expense x_d.
- Total cost of cells x_{gnb} (radio KPIs → bandwidth, number of radio cells)[1]

1 Example: a very high data rate scenario for a large indoor and outdoor NPN. mm-Wave small cells, each of 80–100 m cell range, meaning dozens required per km^2. Part of these cells can be offered by MNO or third party, and part can be taken care of by the enterprise.

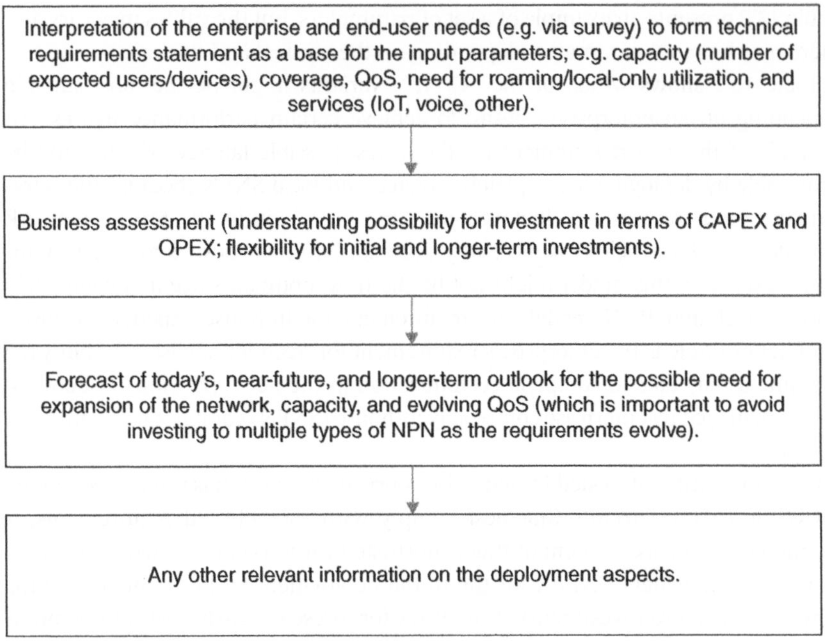

Interpretation of the enterprise and end-user needs (e.g. via survey) to form technical requirements statement as a base for the input parameters; e.g. capacity (number of expected users/devices), coverage, QoS, need for roaming/local-only utilization, and services (IoT, voice, other).

↓

Business assessment (understanding possibility for investment in terms of CAPEX and OPEX; flexibility for initial and longer-term investments).

↓

Forecast of today's, near-future, and longer-term outlook for the possible need for expansion of the network, capacity, and evolving QoS (which is important to avoid investing to multiple types of NPN as the requirements evolve).

↓

Any other relevant information on the deployment aspects.

Figure 13.2 Selection of the most adequate NPN deployment model for further assessment.

- Transmission network x_{tn}
- Spectrum x_s
- Licenses to activate network functions NF x_{nf}
- Applications/services x_a
- Roaming and interconnectivity x_r
- Other variable items x_v

The respective CAPEX can be represented by:

$$\text{CAPEX} = x_{gnb} + x_{tn} + x_s + x_a + x_d + x_v + x_r + \sum_{nf1}^{nfn} x_{nf} \tag{13.1}$$

For the OPEX, the same main components as used in the CAPEX analysis generate expenses, such as licensing fees and electricity consumption, whereas an additional item to be considered in operations is the maintenance cost y_m. The human resource expenses dedicated to operate and maintain the private network can be considered in the value of variable costs y_v.

The respective equation for OPEX is:

$$OPEX = y_m + y_{gnb} + y_{tn} + y_s + y_a + y_d + y_v + y_r + \sum_{nf1}^{nfn} y_{nf} \tag{13.2}$$

The described modeling can be extended to estimate the RoI of the private network, including the business of MNO, enterprise, or third party. The RoI depends on various items, such as:

- Deployment and operational costs.
- Share of ownership of private network components (hardware, software) versus outsourced items (e.g. 5G core that runs in a virtualized environment served by cloud provider) in different deployment scenarios of interest.
- Resulting savings compared to the reference deployment scenario (as an example, enterprises can compare an MNO-hosted scenario against a completely or partially enterprise-owned network).
- Potential earnings for different stakeholders. As an example, an enterprise managing a completely or partially owned private network, either on a shared or own spectrum, could also allow additional users to roam into that network for a fee that depends on the data consumption or time, and there may be value-added services offered too.

The values of the parameters in the CAPEX and OPEX equations depend on more specific circumstances, such as the overall market, vendor pricing strategies, competitive landscape, and scenarios that can change significantly over time. Nevertheless, to test the modeling, the scenarios can be divided into the following categories, considering a network that:

- is completely owned, partially owned, or has completely outsourced ownership
- uses licensed, shared, or unlicensed spectrum
- has no roaming (completely isolated), or has inbound roaming, outbound roaming, or bilateral roaming.

Additional criteria can be assessed for the suitability of the network models based on, e.g.:

- security/level of isolation of the network architecture
- Quality of Service (QoS) and Quality of Experience (QoE)
- latency/responsiveness
- maximum and average data rate
- reliability etc.

The level of compliance of different scenarios can be compared by using numeric values and their importance weighting.

13.7 Example of Modeling

Let us assume an enterprise desires to compare the feasibility of a private network that is either:

- Isolated 5G SNPN with 10×gNBs using unlicensed 5 GHz mm-Wave spectrum, and that has the 5GC NFs implemented on a cloud.
- MNO-hosted NPN on NS dedicated to the enterprise with partially deployed gNBs for PLMN users that are shared with the PN users in the area, complemented with five new additional indoor mm-Wave small cells in the enterprise's operational premises.

In this scenario, there are 300 voice-centric devices with a yearly renewal rate of 1/3 of the total number of devices. The transmission between the gNBs and 5GC is a cost item for the MNO for the already deployed gNBs and for the enterprise for the additional ones. Also, the cost of the activation and operation of the NFs (licenses) are cost items of the operating party. Figure 13.3 depicts an imaginary example of key expense behavior over time using the parameters from Eqs. (13.1) and (13.2) and certain estimated values normalized to the SNPN CAPEX reference at year 0.

As can be seen, in this case, the initial cost of the enterprise's completely own network can be considerably higher than a subscription to an MNO's NS-based PN service due to the investments required in the infrastructure. In this scenario,

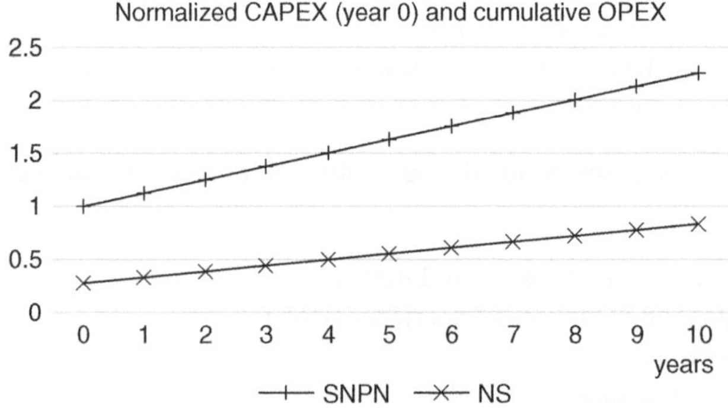

Figure 13.3 Example of the modeling, considering OPEX and CAPEX of enterprise's completely owned and managed SNPN and PN using an MNO network slice.

the OPEX of the SNPN rises faster compared to the dedicated MNO NS due to maintenance and licensing expenses of the enterprise-owned network. Also, the required expertise can have a significant impact on the OPEX along with dedicated human resources taking care of the maintenance, troubleshooting, and upgrades of the enterprise-owned network. Please note that the results can vary significantly depending on each scenario and pricing models between the vendors and operating parties, so there can also be a cross-over between the initially most economic deployment model and more expensive one over a certain time.

13.8 Summary

NPN can serve many verticals and their use cases in a more optimal way than PLMN, to cope with special requirements such as hardened security by isolation, or higher flexibility for network settings adjustment.

Understanding the pros and cons of enterprise-owned versus operators' components, and carrying out techno-economic assessment of enterprises and other entities interested in providing PN services can give a realistic idea of the business impact of each network model of interest.

This chapter presents a means for the assessment of the techno-economic feasibility of NPN models and a fictitious example of the evaluation. For the model to work adequately, insights into realistic OPEX and CAPEX values of the model's parameters are important. Thus, feedback from NPN proof-of-concepts and trials serves to calibrate this modeling and helps identify and focus on the evaluation of the most essential cost items.

Private networks are becoming reality, and they provide a functional basis for many verticals and use cases to cope with special requirements. As this study shows, even a relatively simple model can support the ecosystem and better understand the differences in the business cases related to a variety of private network models.

References

1 GSMA, "Network Slicing: North America's Perspective V1.0," GSMA, 3 August 2021. [Online]. Available: https://www.gsma.com/newsroom/wp-content/uploads//NG.130-White-Paper-Network-Slicing-NA-Perspective-1.pdf. [Accessed 18 December 2022].

2 GSMA, "Generic Network Slice Template V 7.0," GSMA, 17 June 2022. [Online]. Available: https://www.gsma.com/newsroom/wp-content/uploads//NG.116-v7.0.pdf. [Accessed 18 December 2022].

3 GSMA, "Network Slicing: North America's Perspective," GSMA, 2021.

4 J. Penttinen, J. Collin, and J. Pellikka. “On Techno-Economic Optimization of Non-Public Networks for Industrial 5G Applications”. The Nineteenth Advanced International Conference on Telecommunications AICT, Nice, Saint-Laurent-du-Var, France, 26–30 June 2023. [Online]. Available: https://www.iaria.org/conferences2023/filesAICT23/AICT_10003.pdf. [Accessed 27 October 2023].

14

5G-Driven New Business Development

14.1 Introduction

In the Industry 4.0 era, the ability to identify and actively develop new 5G and data-driven business opportunities and to commercialize the developed technology assets is essential for all industries. Therefore, companies are actively looking for new business opportunities related to digital transformation to use to solve customer problems and meet their dynamic needs, for example, related to safety, sustainability, and operational efficiency. The basis of a new business opportunity is to create and manage joint efforts with the customers and ecosystem partners for a thorough understanding of the challenges to be solved. This means that the company and its senior leaders must have the ability to create and capture value with 5G, data, and related industrial applications through a commercialization process for the benefit of their customers. This process is called business opportunity development [1]. Realizing these benefits of technology-based innovations requires an effective commercialization process whereby potential products are generated from ideas and transformed into market-competent products. Further, developing effective commercialization processes is a complex and challenging task for enterprises across industries in the dynamic business environment, in which customer and market requirements are rapidly changing and the life cycles of new products are reducing. This is an especially significant element in digital transformation since many core technologies are evolving so rapidly that companies must match or exceed the pace of change in order to improve their competitiveness and accelerate the twin transition.

Due to these reasons, companies are increasingly jointly building and seeking external technological capabilities, data, and other assets relevant to digital transformation using multi-party ecosystems in order to create value, accelerate

5G Innovations for Industry Transformation: Data-Driven Use Cases, First Edition.
Jari Collin, Jarkko Pellikka, and Jyrki T.J. Penttinen.

Published 2024 by John Wiley & Sons, Inc.

commercialization, and reduce associated risks and costs. For example, building sustainable, digitalization-enabled business models can be very complex and often requires systematic joint efforts among multiple ecosystem partners when it is essential to have a comprehensive understanding of the new opportunities and their implications related to 5G-enabled digitalization. In addition, the shift to digital business models often requires new kinds of capabilities, such as wireless connectivity, data analysis, and software development that challenge industrial companies to be capable of creating value.

New opportunity development can be planned and managed through commercialization, which can be defined as a process that starts with the capability to create value for the customer through idea generation and business opportunity recognition and ends with the maintenance of a marketed product, service, or solution. Thus, a company creates value by converting technological capabilities, data, and innovations into new, or significantly improved, digital solutions, products, and services that satisfy customer and market needs [1, 2]. The commercialization process describes the steps and events that lead to new ways of putting products on the market. A comparison of the concepts of developing a business opportunity and the commercialization process shows that there are several similarities in the chains of events. The different characteristics of a business opportunity developed, enabled by 5G and other modern capabilities, over time and change in the business environment. Therefore, it is essential to understand how value creation can be managed and how it actually takes place across different industry domains in a changing business environment where, for example, digital business models often mean moving from a capital expenditure model, such as the traditional purchase of equipment with add-on repair and maintenance services, to an operating expenses model where the customer pays for an outcome [3].

14.2 Business Opportunity Development and Commercialization to Drive Digital Transformation

From the conceptual point of view, "digital transformation" can be defined as the use of new digital technologies (social media, mobile, analytics, or embedded devices) to enable major business improvements such as enhancing customer experience, digital automation, streamlining operations, or creating new business opportunities, business concepts, and business models [4, 5]. Due to the listed reasons, digital transformation is a critical managerial issue to address in order to reach favorable outcomes. For example, to create value from digital technologies, enterprises need to plan, develop, and deploy new practices to create, deliver, and capture value [6]. In addition, digital transformation involves the ongoing strategic renewal of an organization's ecosystem and a collaborative approach that should be managed using a commercialization process that creates the basis for new digitalization-enabled innovations. Typically, the commercialization process starts with imagining/idea generation and then proceeds through incubation and business concept design toward business model design. The identification and development of business opportunities for digital transformation through this commercialization requires that organizations must be capable of managing and developing both internal (i.e. continuous learning) and external (i.e. ecosystem collaboration) resources, as well as the required capabilities to obtain these resources, which are essential for creating value and enhancing an enterprise's performance. In this approach, the process starts with the ability to create value and progresses to the analysis of the business opportunity such as the identified use cases that can be built using 5G and related capabilities and industrial applications. After this initial analysis, the next steps can include the development of a business concept for a digital business model.

Each phase is associated with several areas, the careful description of which helps the company's strategic planning and decision-making process. Although the importance of commercialization is widely recognized among decision-makers across technology domains, the definition of the term "commercialization" is, in many ways, vague. The term can have a different meaning depending on the context in which it is used. In this chapter, the term "commercialization" is defined as the overall process that is based on the company's ability to create and produce value for the customer using 5G capabilities and with the ecosystem partners. "From this perspective, the commercialization process starts with a techno-market insight and ends with the maintenance of a marketed product, whereby a firm creates economic value by converting knowledge, discoveries, and inventions into new or significantly improved products and services that satisfy customers' needs" needs [1]. This definition encompasses key firm-level activities such as R&D, product development, marketing, and selling, and thus provides a useful basis for exploring the topic in the industrial 5G context.

The success of commercialization is highly important for the success of companies. For example, successful companies can commercialize their solutions two or three times faster than their comparable competitors. This underlines the essential role of data monetization and IoT-related business models, especially when operating in a rapidly changing market [7, 8]. Due to the shortening life cycles of products and services, and in order to maintain competitiveness, companies must be able to develop or replace their products and services constantly in order to secure business and growth. In addition, the emergence of new markets is also influenced by the rapid and accelerating development and spread of 5G-based assets and technologies and the increase in data and its transfer speed as part of the digital revolution [9, 10]. For these reasons, it becomes even more important to succeed in the commercialization of a new product or service and bring a new digital solution to the market as quickly and successfully as possible. In this case, the successful completion of the commercialization process and its speed are key to the company's success, and they are strongly related to the capabilities to create, orchestrate, and develop ecosystem partnerships [8, 9, 11].

As pointed out, rather than relying only on internal resources, firms are increasingly participating in ecosystems and exploiting existing internal and external firm-specific dynamic capabilities and competencies to address rapidly changing business conditions and environments [6]. In particular, dynamic capabilities may enable enterprises to position themselves correctly to make the right products and target the right markets, allowing them to address future customer needs and opportunities. Therefore, the concept that is used during the business opportunity development must also take this characteristic of business dynamics into account. When the business environment and conditions (e.g. customer preferences) change, enterprises need to change the business concept and/or business model to create and capture value that makes the business opportunity development process cyclical and iterative [1]. Therefore, it is likely that an entrepreneur will carry out evaluations several times throughout the different stages of development, and these evaluations can also lead to the identification of new opportunities (strategic options) or adjustments to the initial vision. Companies that are successful in commercialization see it as a strategic and well-planned activity, which is managed and monitored according to the described decision-making and management model, for which measurable goals have been set to monitor progress. Typically, only a few percent of new product and service ideas are successful in the market, which is why targeted investments and expectations cannot be realized according to plans. The basic idea is that, without a market and customers, the value of the new solution is not realized because then no income stream is generated.

The value of a new product or service is therefore linked to the realization of commercialization and the management of the process, which is key to commercial success in the digital era. The following is a list of characteristics that can be found behind successful commercialization processes of new technology-based products, digital business models, and digital services [1, 6, 8, 12–14]:

- Attaching a new product or service idea or technological solution to an identified customer need and a major emerging market opportunity
- Success in introducing a new product or service to potential key customers and influencers
- Close and step-by-step development of a new solution to improve risk management and profitability for a selected group of customers

- Identifying resource needs and their agile acquisition and utilization at different stages of commercialization
- Successful testing and presentation of a new product or service in the customer's real test and/or operating environment
- Successful measures to achieve early market approval and customer commitment, for example, with partners operating in the industrial 5G ecosystem
- Successful identification, selection, and timely utilization of go-to-market channels.

Based on these measures, it is essential to have deep customer and market understanding and the options to create ecosystem partnerships during commercialization. Therefore, their consideration in the first stages of commercialization is very important for companies.

14.3 Main Phases of Business Opportunity Development and Commercialization

In general, the commercialization process of technology-based products and digital services runs from the creation of an idea and the identification of business opportunities, through business concept planning, toward launching the market and developing business models. The decision stage in the process is business planning and further development of the business. This means maintaining the current business while also identifying new strategic options and the activities that are needed to develop these options. The goal of both business opportunity development and the commercialization process is for the company to capture value by transforming observations, existing and new knowledge, abilities, and inventions into new or significantly improved products, services, and/or solutions that satisfy customer needs. In this approach, the organization must be able to manage and target (mobilize) internal and external (ecosystem) resources to achieve profitable business. Practical measures to acquire these resources are particularly important and necessary in all iterative phases of commercialization and business development. In acquiring external resources, building connections with external partners as well as connecting internal and external resources between different stages of the process is called "using bridges." This is an essential part of commercialization, business management, and related decision-making. The main phases of the commercialization process in the industrial 5G context can be described as follows:

Observing the concrete business challenge to be solved using 5G and related capabilities is called business opportunity identification. The starting point for new business is the ability to create added value. To create value, it is essential to be able to identify potential solutions to solve a real customer need. The company acquires information about the characteristics of the business opportunity, based on which it evaluates whether it is worth developing the business opportunity further as part of the commercialization with the selected partners' process. Information is acquired about the attractiveness and sustainability of the market, and decisions are made about the timing of measures, for example, on the market entry. In this context, one's own know-how and the abilities needed for the required asset creation and implementation of the business need to be evaluated. As a result of this business opportunity, a preliminary decision can be made on whether it is worth continuing investment into the business opportunity development or not. This is a crucial phase of digital transformation since the path to transformation is complicated by the need to navigate a fragmented landscape of applications, industrial devices, connectivity, and the number of suppliers. This can lead to longer integration times, higher costs, and an unpredictable return on investment when a detailed analysis of these aspects is essential. For example, in manufacturing and logistics, companies want to use digitalization and automation to accelerate their Industry 4.0 initiatives. To transform their operational processes, they need to connect their assets and generate insights from data while supporting 24/7 operations. The key to success is to combine pervasive, reliable connectivity together with edge computing, digital applications, and industrial devices to meet the demands of everything from sensors and machines to robots and connected workers.

14.3.1 Business Concept Development

As part of the feasibility study of a business opportunity, it is essential to ask what kind of new solution is being developed for the defined use case and who the targeted users of the new digital solution or the potential industrial application are. In addition, it is necessary to determine, for example, how the solution to be developed solves the identified problem, what the concrete benefits are, how success can be measured, and when the benefits can be realized. When describing a 5G-enabled business concept, it is important to take into account the identified problem of the selected industrial domain, the feasibility of the solution in that specific domain, and the business potential, as a main starting point. Industry 4.0 will drive exciting opportunities for the manufacturing and logistics sectors, from digital twins and autonomous mobile robots (AMRs) to remote maintenance and rich group communications. These use cases help to reduce time-to-market, make production lines more flexible, optimize efficiency, and bring the vision of zero safety issues within reach. For example, in manufacturing, the realization of these benefits often requires a common platform that can connect manufacturing processes in real time, enable jointly operating multiple edge computing solutions, and provide high performance, security, and data privacy for mission-critical control and autonomous actions. In addition, the use of 5G and industrial automation needs to be agile to integrate with the existing legacy systems (IT and OT). In all of these examples, the real business benefits and the concrete value-creation elements of 5G capabilities across industries must be identified, including the implementation plan, to ensure value capture. It has been seen that understanding the IoT business models of ecosystem partners is important in the different industry domains for long-term success [15, 16]. For example, digital platforms force established companies to align their existing, traditional business models with the new digital business environments, resulting in business model transformations and digital innovation. In addition, senior leaders need to explore both the opportunities and limitations of digital transformation and the ecosystem of interest. The design space available, the capacities to identify and exploit opportunities, and the radicalness of the idea all influence the likelihood of the emergence of a digital innovation capable of digital disruption.

Although digital business models have been analyzed, a more concrete consideration of the changes in value creation, value offering, and value capture must be analyzed by the decision-makers with a deep understanding of the special characteristics of each industry vertical. For instance, some manufacturers are transforming a traditional business model to a cloud-based digital IoT platform as a multi-sided, industry-wide platform business by making data (e.g. production data to track performance) accessible and useful to their partners and customers. While the manufacturer solely provides the data through defined interfaces, partners can then leverage these data for value capture [10, 16]. In the area of value creation, value offering, and value capture, specific aspects must be considered in Industry 4.0. In addition, it is important to note that the common understanding of the platform and data-based business models in the different industry domains is still relatively limited [16, 17].

14.3.2 Market Launch – Business Model Development with Digital Technologies

In the third phase, the market launch of commercialized digital products and services takes place. When talking about products that have already been commercialized, this stage is called business model development and typically requires strategic ecosystem partnering with the leading companies in their industry vertical to develop wider (sometimes ecosystem-level) offerings. As previously noted, this type of business model development requires joint actions to develop ecosystemic business models with partners across all business sectors and industry verticals to create value [12, 13, 18, 19]. One example is energy and resource-intensive industries such as power utilities, wind farm operators, and oil and gas companies. These industries are currently facing different operational challenges and also an increasing need to develop new business models while reducing carbon emissions and improving safety, security, and productivity across the supply chain at the same time. In addition, digital servitization accelerates even deeper ecosystem partnerships by enabling easier integration and usage of digital assets, industrial applications, and processes [9, 11]. It enables value creation and capture through core digital features, namely monitoring, control, optimization, and autonomous function. The potential for collaborative business models may be significant. Companies are able to enhance their knowledge base and data-driven business growth if they can effectively use their agility and adapt themselves to changes in the market and the current business environment. In addition, companies have the potential to gain from ecosystem collaboration and partnerships due to their ability to use shared knowledge, assets, and resources efficiently through ecosystem collaboration [9, 20]. In the IoT domain, business model innovations tend to cross multiple industries and drive ecosystems in which smart objects facilitate business models and service applications that are incrementally or radically novel in terms of their modularity or architecture [12, 13]. In the process industry, this means that, for example, private 5G and related solutions can deliver pervasive, flexible connectivity that makes it easy to deploy and use new industrial applications and business models and adapt to changing industry needs with low latency, high performance, resiliency, and security capabilities. In addition, data platform-based business models and local and private 5G networks have shown their capabilities to develop new business opportunities and take into account the crucial industry requirements for security, privacy, and vertical-specific and user-specific requirements [17]. These requirements create a very promising basis, for example, for private local 5G networks [21].

Both the creation and the transformation of business models are very challenging for all companies. Organizations are also challenged with managing the complexity of business models around digitized products [22]. Currently, the industry requirements for digital automation, new digital services, applications, and their business models have been very complicated, with a lot of experimentation and many failures [7, 17, 23]. For example, in the process industry, companies may have a mix of siloed wireless technologies to support their operations. These technologies can be inflexible and may have coverage, interference, or quality of service limitations that make it difficult to realize the benefits of digitalization. Therefore, challenges are related to the company's capabilities to create (and capture) value beyond the physical product, which seems to be especially challenging for traditional product companies trying to adapt their existing business models in response to a dynamic business landscape and to tap into 5G and IoT-driven opportunities [7, 21]. In addition, this highlights the importance of business development and commercialization in digital transformation since it is considered to change the dynamics of value creation and value capture through ecosystem collaboration [8].

The output of the commercialization process phase is a business model description. It describes how the company carries out business in Industry 4.0. The business model description includes the value proposition, which is initially already considered in the pre-examination of the business opportunity. In addition, the business model describes the company's operations and customer relationships. The functions described are implementation of sales and distribution and organization and development of customer relationships. Management of collaboration and networks is also part of the business model, as well as finance and cost control. In addition to the economic dimension, the responsible, that is, sustainable business model also includes the ecological and social dimension. This so-called "triple bottom line" model includes the above in addition to the mentioned factors: social services, social impact, community relationships, social channels, ecosystem services, integration of the company's mission (operational idea), environmental impact, ecosystem channels, and ecosystem beneficiaries.

14.3.3 Shared Value Creation and Business Planning

The business planning and development phase includes a compilation of the outputs of the previous phases. In other words, a description of the business opportunity, business concept, and business model form the basis for business planning. The output of this phase is the company's business plan. A business plan is a description of the company, its technology-based products and digital services, and how to make the business profitable. In a changing business environment, an organization's capability to catalyze the emergence and guide the development of an innovation ecosystem can offer an increasing potential and a powerful source of competitive advantage [24]. Working cooperatively with other players, such as private and public organizations, opens up new opportunities in the digital era to use and build complementary assets to drive the organization's objectives further as part of its digital transformation journey [25]. This can be achieved, for example, by using developed novel ways to facilitate resource mobilization to develop new products and services. In order to realize these benefits, orchestration of the IoT-based ecosystem is an essential area to ensure the realization of the value for the ecosystem members.

The plan combines the different aspects together into a functional whole, which helps the company management to take into account the conditions of the operation and avoid misjudgments by paying attention to all business areas. In addition to the description of the first three phases, the business plan presents the company's strategy and financing model. When developing new business opportunities with the ecosystem partners, it is essential to carry out a risk analysis and prepare alternative plans in case iterations are needed during this development phase [2, 25]. This is often the situation in a special industry environment. One example of this is the mining industry, where several mining companies are embracing automation to improve their ability to handle rapid shifts in supply and demand and comply with stringent environmental and safety regulations above and below ground. The automation of critical processes such as drilling, blasting, hauling, and crushing can play a vital role in making mines and pit-to-port operations safer, more efficient, and more productive. Also, broadband data and video communications are essential for an increasingly mobile workforce. However, extreme automation and enriched communications demand robust wireless connectivity beyond the capabilities of Wi-Fi or TETRA solutions.

In this phase, ecosystem orchestration plays a key role in driving joint value capture through iterative joint efforts between the partners. In an industrial 5G context, the term "ecosystem orchestration" can be defined as "the set of deliberate and purposeful actions undertaken by the ecosystem orchestrator or a hub organization to plan, manage and mobilize resources to co-create and co-capture value for the customers through 5G-enabled assets and data-driven innovations." In addition, the orchestration of the innovation ecosystems depends on establishing inter-organizational collaborative practices to facilitate knowledge, asset, and information sharing among the ecosystem members. These practices can be planned and executed by using a systematic approach, including planning and orchestration of the key elements of an ecosystem [9].

In addition, this means that it is comprised of the community of interacting companies and individuals along with their socio-economic environment, where the companies are competing and cooperating by utilizing a common set of core assets related to the interconnection of the physical world of things with the virtual world of the Internet. These assets may be in the form of hardware and software products, platforms, or standards that focus on the connected devices, on the connectivity thereof, on the application services built on top of this connectivity, or on the supporting services needed for the provisioning, assurance, and billing of the application services for further business growth [26]. In addition, from the commercialization point of view, an ecosystem represents the alignment structure of the multilateral set of partners that need to interact in order for a focal value proposition to materialize. This requires a multitude of new collaborations with new and existing actors – for example, new digital infrastructure providers, software and application providers, connectivity providers, and specialized small and medium-sized enterprises (SMEs) – as well as existing local sales and service partners (e.g. integrators and distributors) and competitors who need to align for a focal value proposition to materialize.

14.3.4 Digital Technologies and Strategic Options

Today, a major proportion of industrial assets, machines, and field workers are not connected in any way, which means there are limited opportunities to use, for example, real-time data for operational insights. In addition, IT and operational technologies (OT) are often incompatible because of complex industrial protocols, certifications, and skills gaps. All of this makes it more challenging for enterprises to embrace Industry 4.0 and adopt new applications and devices. To succeed with digital transformation, an ability to capture and act on data, often in near real time, is required. Therefore, it is essential that data from the different industrial systems and IoT sensors can be processed by edge computing systems enabled by artificial intelligence (AI) and machine learning (ML) to allow the modeling and analysis of such data for the continuous improvement of industrial operations and processes. The results can be then used to make effective and data-based decisions, for example, to optimize processes and to realize efficiencies for industrial automation. After the strategic initiatives have started, the identification of further strategic options can be defined [1, 26], thus starting the process again from step 1, i.e. the feasibility study of the business opportunity.

The main phases described in the model are connected by four joins ("bridges"), which are intended to ensure creation of value for customers at different stages, guaranteeing the progress of a digital solution development that meets expectations and a robust introduction of new products and services with future market potential outside the selected customer groups [1, 13, 15]. The most important task of the bridge between the first and second stages is to ensure that there is renewed interest in the product or service among customers. This bridge also exists to verify that the main stakeholders are convinced that it is worth proceeding to the conceptualization phase in the commercialization process and that the operational plan for it can be accepted. The second bridge is located between the planning and market launch phases of the business concept. The most important function of this bridge is to ensure that from business concept to the market launch, the measures needed to develop the business model and the resources for their implementation are at the company's disposal. A particularly important role is played by resources outside the company, ensuring availability and strengthening the commitment of stakeholders, especially selected main customer groups as users of the product or service.

In addition to this, the goal should also be to find customers to support the launch of a broader business concept. This can happen, for example, with recommendations made by the customer to their own stakeholders ("co-marketing") and with market communication supporting the launch. Placed between the market launch and business planning and development phases, the main goal of the bridge is to reach a wider market (including selected customer groups and/or outside the market areas) for approval for the newly launched solution. In addition, one of the main advantages is to adapt the operation and solution's operating environment so that the new solution will give its users as much value as possible. For example, in ecosystems, organizations can jointly develop business opportunities, opened up by 5G technologies, by strongly engaging first-time users who have a significant impact on achieving wider market acceptance of the new technology. The bridge between the stages "business planning and development and development of strategic opportunities" includes decisions about the resources needed in the last stage and other measures. These measures are focused on and around problem-solving for the continuous improvement of the structured business as part of the company's business management, market situation, and anticipated changes.

From a managerial perspective, there are several factors that can be used to support and develop the commercialization process in enterprises across industries as part of digital transformation. As indicated earlier, digital technologies and data have changed the way business-to-business companies plan and carry out their business. The amount of business-relevant data has continuously increased over the past few years, and new analytical approaches combined with computational assets have created new opportunities to turn data into meaningful insights. As part of digital transformation, enterprises collect, transport, and use data. It is essential to note that commercial value cannot be extracted from data if the company does not generate or have access to that data, and if that data is not transmitted and stored appropriately [14]. Therefore, without the combination of data, data-usage rights, connectivity, and real-time analytics enabled by ML/AI capabilities, only limited value can be realized. Therefore, practicalities to cover all aspects (commercial, legal, and technical) of data must be defined and implemented as part of a commercialization process to drive digital transformation. For example, [27] underlined the differences between the concepts "digital" and "digitized," where the former concerns digital value propositions in the marketplace and the latter relates to the transition from analog data to digital data, which streamlines existing processes.

The move toward Industry 4.0 is complex and involves the interplay of multiple interacting process instances, raising significant challenges for managers attempting to modernize their businesses. The results indicate that senior leaders recognize that commercialization is a vital, integral part of their digital transformation and business opportunity development with the ecosystem partners. Enterprises must be capable of developing new leadership structures and skills to drive digital transformation and utilization of digital technologies [27, 28]. In the dynamic business environment, organizations should adapt their current conditions to the new developments in technology. From a broader perspective, when customers or stakeholders explore the facilitation of the new technological development before the organization, this may result in changing the preferences of the customers and transferring to the other organization, which can follow the new updates in the digital era [28]. Thus, an appropriate business model needs to become part of the new dominant logic for managing new opportunity development and commercialization. In addition, attention should be paid to value capture with digital technologies to help enterprises better estimate the business potential of the different use cases that are driven by new 5G opportunities. It is also important to note that technology-related decisions automatically bring about changes in, for example, the company's marketing capability. Thus, from the managerial perspective, for effective management of commercialization, the technology already in use (and, as far as possible, being developed by competitors) needs to be monitored to acquire information required to foresee future changes and modifications that need to be made in order to build sustainable business through digital transformation. This means that enterprises require innovations to meet the changing needs and preferences of the markets as a result of rapid digital technology developments. In addition to the vision of the leaders of the organizations, digitally literate leaders support and motivate the employees and help them show innovative work behavior. Organizations should pay attention to digital transformation in the twenty-first century and develop business models based on innovative approaches [13, 28].

References

1 M. Virtanen and J. Pellikka, "Integrating the opportunity development and commercialisation process," *International Journal of Business and Globalisation*, vol. 20, no. 4, pp. 479–496, 2018.

2 J. Pellikka, M. Kajanus and M. Seppänen, "Open innovation adoption practices and evaluation methods in the global process industry," in *Open Innovation: A Multifaceted Perspective: Part I*, 2016, pp. 181–205.

3 D. Sjödin, V. Parida, M. Jovanovic and I. Visnjic, "Value creation and value capture alignment in business model innovation: A process view on outcome-based business models.," *Journal of Product Innovation Management*, vol. 37, no. 2, pp. 158–183, 2020.

4 M. Fitzgerald, N. Kruschwitz, D. Bonnet and M. Welch, "Embracing digital technology: A new strategic imperative," *MIT Sloan Management Review*, vol. 55, no. 2, 2014.

5 T. Hess, C. Matt, A. Benlian and F. Wiesböck, "Options for formulating a digital transformation strategy," *MIS Quarterly Executive*, vol. 15, no. 2, pp. 123–139, 2016.

6 K. Warner and M. Wäger, "Building dynamic capabilities for digital transformation: An ongoing process of strategic renewal," *Long range planning*, vol. 52, no. 3, pp. 326–349, 2019.

7 H. Chan, "Internet of things business models," *Journal of Service Science and Management*, vol. 8, no. 4, p. 552, 2015.

8 S. Leminen, M. Rajahonka, M. Westerlund and R. Wendelin, "The future of the Internet of Things: Toward heterarchical ecosystems and service business models," *Journal of Business & Industrial Marketing*, vol. 33, no. 6, pp. 749–767, 2018.

9 J. Pellikka and T. Ali-Vehmas, "Managing innovation ecosystems to create and capture value in ICT industries," *Technology Innovation Management Review*, vol. 6, no. 10, pp. 17–24, 2016.

10 B. Stahl, B. Häckel, D. Leuthe and C. Ritter, "Data or business first? – Manufacturers' transformation toward data-driven business models," *Schmalenbach Journal of Business Research*, pp. 1–41, 2023.

11 S. Leminen, M. Rajahonka, R. Wendelin, M. Westerlund and A. Nyström, "Autonomous vehicle solutions and their digital servitization business models," *Technological Forecasting and Social Change*, vol. 185, p. 122070, 2022.

12 S. Leminen, M. Rajahonka, R. Wendelin and M. Westerlund, "Industrial internet of things business models in the machine-to-machine context," *Industrial Marketing Management*, vol. 84, pp. 298–311, 2020.

13 P. Ahokangas, M. Matinmikko-Blue, S. Yrjölä, V. Seppänen, H. Hämmäinen, R. Jurva and M. Latva-aho, "Business models for local 5G micro operators," *IEEE Transactions on Cognitive Communications and Networking*, vol. 5, no. 3, pp. 730–740, 2019.

14 T. Ritter and C. Pedersen, "Digitization capability and the digitalization of business models in business-to-business firms: Past, present, and future," *Industrial Marketing Management*, vol. 86, pp. 180–190, 2020.

15 A. Leiting, L. De Cuyper and C. Kauffmann, "The Internet of Things and the case of Bosch: Changing business models while staying true to yourself," *Technovation*, vol. 118, p. 102497, 2022.

16 T. Şimşek, M. Öner, Ö. Kunday and G. Olcay, "A journey towards a digital platform business model: A case study in a global tech-company," *Technological Forecasting and Social Change*, vol. 175, p. 121372.

17 M. Mesquita, A. Simões and V. Teles, "The role of digitalization, servitization and innovation ecosystem actors in boosting business model innovation – A literature review," *Innovations in Industrial Engineering II*, pp. 114–127, 2022.

18 T. Chin, Y. Shi, S. Singh, G. Agbanyo and A. Ferraris, "Leveraging blockchain technology for green innovation in ecosystem-based business models: A dynamic capability of values appropriation," *Technological Forecasting and Social Change*, vol. 183, p. 121908, 2022.

19 M. Palmié, L. Miehé, P. Oghazi, V. Parida and J. Wincent, "The evolution of the digital service ecosystem and digital business model innovation in retail: The emergence of meta-ecosystems and the value of physical interactions," *Technological Forecasting and Social Change*, vol. 177, p. 121496, 2022.

20 S. Gbadegeshin, "The effect of digitalization on the commercialization process of high-Technology companies in the life sciences industry," *Technology Innovation Management Review*, vol. 9, no. 1, pp. 49–63, 2019.

21 M. Matinmikko, M. Latva-Aho, P. Ahokangas, S. Yrjölä and T. Koivumäki, "Micro operators to boost local service delivery in 5G," *Wireless Personal Communications*, vol. 95, pp. 69–82, 2017.

22 M. Wen, Q. Li, K. Kim, D. López-Pérez, O. Dobre, H. Poor, P. Popovski and T. Tsiftsis, "Private 5G networks: Concepts, architectures, and research landscape," *IEEE Journal of Selected Topics in Signal Processing*, vol. 16, no. 1, pp. 7–25, 2021.

23 A. Klein, F. Pacheco and R. Righi, "Internet of things-based products/services: Process and challenges on developing the business models," *JISTEM-Journal of Information Systems and Technology Management*, vol. 14, pp. 439–461, 2017.

24 M. Berawi, N. Suwartha, M. Asvial, R. Harwahyu, M. Suryanegara, E. Setiawan, I. Surjandari, T. Zagloel and I. Maknun, "Digital innovation: Creating competitive advantages," *International Journal of Technology*, vol. 11, no. 6, pp. 1076–1080, 2020.

25 P. Friess and R. Riemenschneider, "New horizons for the Internet of Things in Europe," in Vermesan, O. and Friess, P. (eds.) *Building the Hyperconnected Society-Internet of Things Research and Innovation Value Chains, Ecosystems and Markets*, River Publishers, 2022, pp. 5–13.

26 R. Adner, "Ecosystem as structure: An actionable construct for strategy," *Journal of Management*, vol. 43, no. 1, pp. 39–58, 2017.

27 J. Ross, C. Beath and M. Mocker, *Designed for Digital: How to Architect Your Business for Sustained Success*, MIT Press, 2019.

28 M. Wagner, F. Heil, L. Hellweg and D. Schmedt, "Working in the digital age: Not an easy but a thrilling one for organizations, leaders and employees," in *Future Telco Successful Positioning of Network Operators in the Digital*, pp. 395–410, 2019.

15

Next Steps Toward the Industrial Metaverse and 6G

15.1 Introduction

The term "metaverse" is generally defined as a collective virtual online environment created by the fusion of physical and digital reality. It is expected that the metaverse can potentially solve the problem of creating a fully immersive and interconnected virtual world that can be experienced by people in a way that is similar to the physical world [1]. It aims to create a shared virtual space where people can interact with each other, engage in commerce, and access a range of digital services and experiences. The metaverse is seen as a potential solution to the limitations of the physical world, such as geographical barriers, accessibility issues, and environmental concerns, i.e. it may enable better democratized interactions and a digital economy with virtual asset transfer in a decentralized manner. From this perspective, the metaverse, for example, creates a real–virtual continuum that allows the introduction of virtual elements into real environments through augmented reality (AR) or placing real actions in virtual environments through augmented virtuality (AV). The experience in metaverses is situated in this continuum of mixed reality (MR) that helps to develop new opportunities for design, usability, and representation in physical–virtual worlds [2]. It provides a synchronous and persistent experience for unlimited users.

The metaverse is initially being driven by enterprises that are focusing on business sectors such as gaming, social media, and advertising. However, it is already increasingly supporting industrial use cases, e.g. through the digital twin concept that has been introduced previously. It is often viewed as the next stage of the internet (Web 3.0), and it is assumed that the metaverse will continue to evolve over the next few years, with a clearer scope within five years. The foundational building blocks of the metaverse are not new, but they will create several new

5G Innovations for Industry Transformation: Data-Driven Use Cases, First Edition.
Jari Collin, Jarkko Pellikka, and Jyrki T.J. Penttinen.

Published 2024 by John Wiley & Sons, Inc.

requirements for the current 5G networks to create value in the metaverse era since they will likely impact multiple industries through new business models and ecosystem collaboration as well as through transformations on how consumers and enterprises communicate, perform, and consume data [3, 4]. Therefore, in Industry 4.0, we may see much more demanding requirements for latency, jitter, reliability, efficiency, and sensing that are expected to be driven by specialized high-performance sub-networks and dedicated networks deployed locally [5, 6]. In addition, as indicated earlier, the importance of energy efficiency, cybersecurity, and other new service requirements from the different domains will create new challenges to be solved for the industry.

For example, international policy framework is setting new requirements for companies and cities around the carbon footprint that will accelerate green transition across industrial operations. In many industries, sustainability of digitalization must be taken into account because, for example, mobile network power consumption is linked to exponential data traffic growth. Energy consumption accounted for between 15% and 40% of teleoperators' OPEX in 2021 and is expected to increase heavily in 2022 and onwards [7]. The energy costs associated with running the world's mobile networks are expected to exceed 24 B€ annually due to the ongoing energy crisis and inflation. The estimated annual increase in energy costs will be 8–12% [7]. At the same time, the cyberattacks at the global level have increased by 38% in 2022, compared to 2021 [8]. In addition, global cybercrime costs are estimated to grow by 15% annually over the coming years, reaching 10 trillion euros annually by 2025 [8]. These trends, together with the new service requirements, create the basis for an industrial metaverse that will be described in the next section.

Compared to other metaverse categories (i.e. consumer and enterprise metaverses), the industrial metaverse is expected to provide the following key enablers and enhancements for the industries [5, 9, 10]: (1) Private networks that further drive and enable Industry 4.0, (2) Industrial automation, (3) Digital twins for production optimization, and (4) 3D maps for autonomous mobile robots. These capabilities can accelerate new business opportunity development in the future, and it has been estimated that, in particular, the industrial metaverse is projected to lead the way in commercialization. For example, [11] forecast a US$100 billion industrial metaverse market by 2030, with massive revenue potential from digital twins, extended reality applications, and more. Used for everything from creating ecosystems when planning a new city to working out iterations of manufacturing processes, digital twins were first proposed in 2002 and later became a vital technology when the fourth industrial revolution (Industry 4.0) accelerated automation and digitization across industries [11]. Revenues from industrial digital twin and simulation and industrial extended reality (XR) will hit US$22.73 billion by 2025 as organizations use Industry 4.0

tools such as artificial intelligence (AI), machine learning (ML), edge computing, and XR to accelerate digital transformation [11]. However, although the potential of the metaverse has already been seen, there are many building blocks that must be created to provide assets needed for value capture. These fundamentals will be briefly described in the next section.

15.2 Fundamentals of Industrial Metaverse

It has been estimated that industrial application-driven "industrial metaverse" will grow into a US$100 billion market by 2030 according to [11]. Some selected industrial verticals such as manufacturing, process industry, transportation, and utilities have started their investments in industrial metaverse applications to improve their productivity, accelerate their green transition through VR/AR/MR and 5G technologies supported by ML/AI capabilities, and create additional value for their customers. The industrial metaverse also has immense potential to enhance planning, testing, and operations, leading to better decision-making and improved worker safety [12–14]. It is expected that the speed of this development will increase across industry verticals, which will create new business opportunities for many organizations.

However, this will require increased cooperation between network function vendors, network service providers, application service providers, and hyperscale companies within an evolved ecosystem to revolutionize networks and related applications for industrial metaverse I, where physical and virtual worlds will merge. Therefore, concepts such as "Networks of Networks" will be needed to improve some of the current limitations on coverage and capacity that are needed for future industrial metaverse applications. In addition, more intense collaborative integration and extension of traditional networks with other types of networks, such as non-terrestrial low earth orbit (LEO)-based networks, localized mesh networks, or even decentralized networks, are expected to be one future avenue to develop new capabilities for the industrial metaverse use cases. Based on these expectations, it has been estimated that, for example, models and concepts on Network-as-a-Service (NaaS) will increase their level of maturity by 2030, and a significant proportion of models will be based on a service, enabled by boundary resources such as APIs (e.g. Network as Code) and AI/ML-driven automation [5, 15].

As we have pointed out, 5G networks in the industrial context have multiple capabilities and functions to meet the requirements of different industry verticals and, more generally, of Industry 4.0 applications such as remote inspection using drones, public safety, automated terrestrial vehicles, supply chains driven by digital twins, adaptable manufacturing systems, and maintenance utilizing MR [15, 16].

This is where a network-powered ecosystem will play an essential role. A network-powered ecosystem that involves communications service providers (CSPs), app developers, and enterprises is crucial for driving new revenue from Industry 4.0. All stakeholders in this ecosystem aim to benefit: CSPs by providing developers access to their network properties, developers by delivering new application experiences, and enterprises by offering new products and services. However, developers and industrial application providers are not typically connectivity experts, and it is essential to provide them access to a wide variety of network capabilities through network APIs. In addition, the current solutions limit what developers can do with the network as they focus only on a narrow set of communication service APIs, whereas developers will need access to a much broader set of network capabilities.

The network-as-code approach underlines the increasing importance of application developers to the emerging metaverse opportunities. In the network-as-code approach, third-party application developers can more effectively utilize network capabilities for their own skillset and solution development, which enriches the ecosystem's capabilities to create wider value. For example, communication platform as a service (CPaaS) can create secure access to messaging, voice, and video capabilities through programmability in various digital ecosystem applications [17]. Programmable networks, for example, involve developers within enterprises, CSPs, and third-party independent developers [6]. This provides a win-win proposition for the developers, the CSPs, and their enterprise and consumer customers. Through these network APIs, application developers can potentially leverage many network capabilities globally, and CSPs can enable their networks to be accessible to industrial application developers for different verticals and markets in which they operate [15]. Developers working on industrial and enterprise applications are aiming to enhance business efficiency, customer engagement, sustainability, and safety by optimizing the building, visualization, operation, and management of physical objects and processes.

From the sustainability point of view, 5G-Advanced will be a milestone in the energy-efficiency evolution of networking, as 3GPP has made minimizing power consumption a major focus of mobile standards from Release 18 onwards. In addition, this means that 5G-Advanced is looking at new ways of minimizing energy consumption in the network, and 3GPP is developing a new energy consumption model for 5G-Advanced, focusing on the base station and radio access network, which together use the vast majority of the network's electricity. These activities accelerate the creation of new techniques for minimizing power use on both the uplink and downlink, as well as when the network is dynamically transmitting data or passively waiting in standby.

These new network capabilities can help to solve some of the current new service requirements of metaverse use cases across industries. Since many use cases for the industrial metaverse may involve a significant number of IoT devices that create data, it is important that the latency requirements can be met for the use cases. For example, haptic-based applications and the exponential growth of the volume of data and models force the development of new network capabilities for the future use of industrial 5G. Other key elements in the metaverse are security aspects that must be taken into account in all the use scenarios of the industrial metaverse. For example, industrial immersive experiences can be based on wearable devices such as VR and mobile headsets, and the collected data from these devices will be transmitted to the cloud. The new application scenarios in the near future, the multiple stakeholders within each scenario, and the higher data volumes raise the need for novel cybersecurity solutions. Recently, the cybersecurity threat landscape has become wider due to the increased digitalization of services, the increase in virtualization and slicing of networks, as well as the increase in advanced cyberattacks. Following recent advances in computing power, AI in cybersecurity is now becoming a reality. In addition, a significant part of currently used encryption, which secures critical communications and infrastructures, might become instantly breakable when quantum computing becomes available.

As indicated above, these new requirements also open up future opportunities for new business and service models [5, 9]. These new network-based business models and emerging technologies are taking industrial 5G to the next level and driving further productivity and sustainability improvements. It is expected that the future development of multiple technology domains will contribute to the data-driven 5G-Advanced industrial use cases, such as digital twins, robotic production lines, logistics hubs, drone highways, smart cities, and virtual workplaces, reaching a higher level of virtual reality. For example, a digital twin is a virtual entity that digitally recreates a physical entity. A digital twin collects various data for the physical model in the real world through simulation technology and maps a digital body exactly like the object on the screen through these data. Digital twins can observe the operation of the digital entity in real time, monitor various processes and operational functions, realize further simulation of the virtual world through data accumulation combined with AI, and then create the feedback loop back to the physical world [10]. With the help of historical data, real-time data, and algorithm models, it is a technical means of simulating, verifying, predicting, and controlling the entire life cycle process of a physical entity. Although the metaverse and digital twins focus on the connection and interaction between the real and virtual worlds, the essential difference between the two is that the metaverse is directly oriented toward people, whereas digital twins are oriented toward things [18, 19].

The concept of the metaverse represents the development direction of the next generation of the internet and, along with digital twins, the two major technology systems complement each other and will lead the fourth industrial revolution driven by technologies including AI/ML, haptics, motion tracking, volumetric video streaming, photorealistic rendering, hyper-accurate positioning, and virtual reality [15]. As pointed out, many industrial metaverse use cases such as digital twins, immersive AR/VR, and remote collaboration have very different requirements for optimal connectivity, consumability, and performance. This means that there is a need for network transformation in multiple aspects, in order to meet these new demands, as well as to create the capabilities to add value around these new service opportunities [5, 9]. Some of this transformation can be realized through new paradigms in building and integrating future networks for industry purposes through new capabilities, as described previously by [2, 12, 19].

Domain-specific customization and optimization that enable dedicated customer-specific networks and related new business models provide optimal connectivity and ecosystem-based industrial applications as a premium service. In addition, this would also support the widest range of industrial metaverse use cases as well as dynamic capacity demand fluctuations, including the capability for rapid setup and take-down of localized or private networks. Many of the listed industrial metaverse use cases also need network resources that can be deployed dynamically to provide the best user experience while maximizing resource utilization and enabling networks to adapt better to industrial needs autonomously using the most advanced AI/ML capabilities. On top of these generic industrial needs, from the metaverse point of view, security and sustainability must be built into the future network capabilities, for example, to allow the network to both extend its handprint and dramatically reduce its footprint while data traffic continues to grow significantly in the future [17, 20].

From industrial point of view, the metaverse can potentially give enterprises new options to cut costs, act more sustainably, improve the work experience, and accelerate operations by moving expensive, time-consuming activities into fully virtual environments. Within these spaces, enterprises and public sector agencies can test and optimize their systems, processes, and infrastructure to identify and address issues before committing time and resources to the real thing. Companies are currently using 5G networks to send and receive massive amounts of data for their industrial metaverse initiatives. The 6G mobile system standard, which is referred to as a platform for the metaverse, is likely to make the industrial metaverse even more attractive [5]. The 6G standard is currently under development and is expected to expand the framework of cellular standards to use XR in business and gaming environments. More importantly, 6G is expected to be more energy efficient than current cellular networks, and, thus, sustainability is expected to become a major component of this standard [5].

The industrial metaverse combines physical–digital fusion and human augmentation for industrial applications and contains digital representations of physical industrial environments, systems, assets, and spaces that people can control, communicate, and interact with. It has been expected that in industrial and enterprise segments, the metaverse could drive up to 9× higher bandwidth data consumption by 2030. Therefore, the networks must transform to meet these new requirements and must also evolve to provide key enablers for realizing the value presented in the new opportunities emerging in the different metaverse categories. For example, in order to accelerate digital transformation across industries, requirements on low latency, jitter, reliability, efficiency, and sensing will be managed using specialized high-performance sub-networks and locally deployed private networks.

Due to these requirements, especially from the industry domains, the metaverse is expected to be a complex interconnection of services from multiple stakeholders. This interconnectivity complexity underlines the critical role of future networks and network services. For the metaverse to become accepted and adopted at scale across industries, the experience must address latency to avoid slow response or jittery motion [4]. Both network and application latency will be key. For example, future wireless networks will require higher uplink and downlink capacities to support the volumes of connected devices and the data required to deliver VR/AR/XR in a consumable way. The industrial metaverse use cases can have a much greater diversity of items, assets, and environments that are rendered on an as-needed basis, requiring an abundance of cloud-streamed data. In addition, sufficient edge infrastructure will be required for localized application and analytical computing needs, and therefore, for example, the current capabilities for AI/ML-enabled edge and digital service orchestration and app workload management infrastructure must be developed. Application assurance will be challenging, driving the importance of service orchestration, so the component services must be orchestrated efficiently to meet the time-sensitive needs of mission-critical processes. A robust foundation built on multi-access connectivity and high-speed computing functions is essential to take full advantage of the industrial metaverse as it evolves. In addition, secure, ultra-fast networks with low latency and high reliability are required to power the industrial metaverse and keep mission-critical applications online. The need for near-zero latency makes edge computing a must for bringing enterprise applications closer to data sources, enabling greater and faster processing.

In particular, for mission-critical industrial applications, the metaverse will require low latency, massive machine communications, and high reliability, in addition to fast network speeds [4, 13, 21]. Edge computing is another must-have because of the requirement for almost zero latency. Thus, decentralized local edge data centers close to users will be needed for people to interact with one another and use devices to access the metaverse. In addition, AI-based motion control and the use of AI in analyzing images and videos with computer vision can help us better understand the environment and act in the best possible way. In order to create and combine these key building blocks for the metaverse, an ecosystem approach is needed. An ecosystem of partners, technology and network providers, data producers and owners, and application developers will contribute to these building blocks. Collectively, they facilitate a digital marketplace and lead to new and unprecedented levels of innovation, creativity, and agile and collaborative service creation. As with any innovative technology, security is paramount, especially because cyberattacks have surged in recent years, with criminals employing increasingly sophisticated technology such as ML/AI. Therefore, keeping people's identities secure and protecting the data shared within virtual collaboration will also be integral, especially across a decentralized ecosystem of stakeholders who may not have preestablished business and relationships with one another. From the network requirements perspective of the metaverse, radically enhanced performance through specialized dedicated networks will be needed because of the anticipated extreme capacity of the metaverse in both uplink and downlink traffic and optimization for a wider range of customer-specific needs (e.g. industrial). In addition, future networks (5G-Advanced and beyond) must enable functionalities such as zero-touch management and orchestration using AI/ML-driven capabilities and security and energy efficiency features, including all aspects of the network, designed as core requirements.

For example, AI/ML capabilities help industrial applications digitally learn from real-time experience to enable modeling, prediction, and automation. In addition, AI/ML-enabled 5G cybersecurity allows industry to protect networks through scalable controls, well-integrated Defense in Depth systems, and end-to-end security orchestration and automation [17, 20]. In order to create a solid basis for future industrial metaverse assets and mission-critical applications, digital industrial platforms will play an even more important role in the future [19, 22]. For example, these platforms enable collection and usage of industrial IoT data and leverage this data for the creation of smart applications and services with ecosystem partners such as digital service developers, system integrators, and other business partners. Data from the various sources must also be synchronized across all elements, as real-time applications require a single source of truth, even if distributed across multiple edge cloud environments. By definition, the decentralized and service chain nature of the metaverse includes elements outside of the control of the experience provider. Networks must intelligently and autonomously provide insights and recommend actions based on operational data from an array of infrastructure and multi-vendor platforms to deliver the committed service levels [3, 13].

5G-related technologies will increasingly be able to address metaverse infrastructure needs. For example, in future networks, digital twin capabilities will enable the maintenance and monitoring of network operations in real time and predict maintenance needs and potential downtime in advance. Therefore, they can drive productivity by reducing disruptions in production. Digital transformation and Industry 4.0 need data to capture value and, therefore, a digital twin can be used to: (1) Understand what has happened in the network, what is happening now, and what will happen when new use cases are deployed, (2) Identify capacity, performance and/or coverage limitations before the introduction of automated solutions and machines at the premises, and (3) Use automated troubleshooting and fast resolution of the identified challenges without having to dispatch workers to local campuses.

Sectors such as manufacturing and logistics are expected to drive industrial metaverse growth. Early pioneers of digitalization, these sectors have been using enabling technologies such as AI/ML, XR, and digital twins for the last few years. The success of their efforts will accelerate the widespread adoption of the industrial metaverse across other industries and sectors, including railways, power utilities, and public safety. However, it is important to note that new business opportunity development and commercialization call for ecosystem partnerships. It has already seen a shift toward a multi-party value ecosystem, centered around and leveraging the new network capabilities and NaaS propositions in ecosystem-based business models. No single player will be able to provide, or own, all the elements of the future metaverse solutions. Many existing and new players will be able to innovate around software, devices, and new concepts in the metaverse, and it is expected that well-planned and orchestrated ecosystems can create a basis for concrete asset building, e.g. for the mission-critical applications and the platform where the ecosystem partners can collaborate, co-innovate, and partner. In addition, it is expected that the role of the developer will become increasingly important in a world that rewards speed of commercialization and capabilities to develop new business opportunities at the global scale effectively. However, decision-makers and senior leaders must be aware of the following areas of the 5G-enabled business models in the future to create and capture value for the ecosystem and society in a sustainable way when moving toward the industrial metaverse.

First, ecosystem partners must have a vision, a core purpose, and key objectives that remain relatively stable while the strategies and practices continuously adapt to a changing environment and accelerate new business opportunity building and commercialization. It is essential that their vision creates a fundamental, ambitious sense of purpose, one to be pursued over many years. Powerful visions clearly indicate the long-term approach to how the ecosystem and related ecosystem-based business models create and capture value for the member organizations in the digital–virtual worlds. This is because a vision has no power to inspire ecosystem members or attract new members to join unless it offers a view of a sustainable future. In addition, based on the articulated ecosystem vision, the decision-makers can create an ecosystem strategy, which is key to an ecosystem's operations [23–25]. Finally, once the ecosystem has developed a vision of what market it wants to enter and with what offerings, it comes up with a tentative agreement on the performance expectations that enable it to reach the set objectives.

15.2.1 Business Environment in the Industrial Metaverse

Ecosystems can enhance their performance in a dynamic business environment by focusing on their dynamic capabilities [23, 26]. Dynamic capabilities can help ecosystems adapt to emerging changes in their business environments. In addition, they can help to identify and develop new innovation and business opportunities and, in general, to maintain competitiveness through enhancing, combining, protecting, and, when necessary, reconfiguring the organization's intangible and tangible assets. As previously pointed out, the industrial and business environments are constantly forming and transforming through exploration, mobilization, and stabilization. Therefore, a broad environmental study of emerging changes and opportunities is essential in order to know what is currently happening and what is going to happen [25]. From this perspective, ecosystems must develop their capabilities to adapt and, beyond that, be complemented by sufficiently agile structures and processes so that the ecosystem and the ecosystem members can also effectively adapt relevant operations. Building these types of capabilities also requires decentralized decision-making, a collaborative organizational culture in the ecosystem, and a shared vision among the ecosystem members to develop new business opportunities for industrial metaverse use cases.

15.2.2 Ecosystem Management and Governance

Although innovation ecosystems have been increasingly used by companies to foster innovation through collaboration, there are still challenges regarding how to orchestrate technology-based ecosystems successfully [26, 27]. It has been noted that one main challenge is to orchestrate a network of actors, assets, data, and resources effectively in the ecosystems. In addition, assets and resources must be orchestrated by a strong entity willing to take the lead [27, 28]. Therefore, ecosystem management and facilitation are crucial elements in order to realize the potential benefits of an ecosystem. Management and facilitation create a basis for good governance and structured management practices [25]. In other words, it is crucial that ecosystems are managed and facilitated using structured methods based, for example, on the predefined and articulated governance model. Some ecosystems have clear, possibly de jure defined standards, especially if they have a large number of members, for example, on the common digital platform.

15.2.3 Capabilities and Complementary Assets

Previous studies have clearly indicated an essential role in the innovation ecosystem for jointly creating, using, and further developing knowledge, capabilities, data-based resources, and complementary assets [27, 29]. In addition, ecosystems need: (1) capabilities to manage data-based assets effectively (both internally and externally), (2) good-quality knowledge-based assets, and (3) the successful application of these assets to fulfill the organization's strategic objectives through ecosystem management. Therefore, a company's resources should not only be valuable, rare, and inimitable to facilitate superior performance, but the company must also have an appropriate strategy, organization, and processes in place to take advantage of the knowledge-based resources and data within the ecosystem. In addition, it has been previously noted that the increasing role of data and knowledge has become one of the primary wealth-creating assets enabling innovation, business opportunity development, and commercialization. It has been previously highlighted as playing an essential role in the data-based view (DBV) in the ecosystem context [27]. Digitalization of society is a key opportunity for techno-entrepreneurs and start-ups to commercialize innovations successfully for the emerging industrial metaverse use cases. Therefore, accessing and effectively utilizing complementary data and knowledge-related assets in an ecosystem is a critical success factor. A large majority of industry sectors and new business opportunities (e.g. IoT and digital healthcare) are generated by an ecology of private, public, and nonprofit organizations, all often involved in innovation ecosystems. In order to stay relevant in the business, all ecosystem stakeholders must strategize and continuously align both the inbound and outbound data and knowledge flows using predetermined practices, e.g. through effective boundary resources, as also noted in the case studies.

Furthermore, previous studies have shown that platforms can accelerate the value recognition function of absorptive capacity and, therefore, accelerate further diffusion of knowledge, data, and knowledge acquisition and co-development among the industrial metaverse ecosystem members. These activities can also enable start-ups and small and medium-sized enterprises to deepen their specialization while further developing their business opportunities through business concepts, business models, market launching, and business planning. The final vital element in the ecosystem is the need for at least one industry leader company or a "keystone" company. Their role is to ensure the continuous improvement of the ecosystem and engage new innovative start-ups to join the ecosystem and create offerings that are compatible with the expectations of other ecosystem stakeholders, including end-users. This role may coincide with the roles and activities taken on by the orchestrator(s) within the innovation ecosystem.

15.2.4 Monitoring Impact and Change Management

Innovation ecosystems typically consider a complex and dynamic structure as they bring multilevel perspectives and capture the complex relationships that are formed between multiple actors such as large companies, SMEs, start-ups, universities, government, NGOs, citizens, regional communities, infrastructure, customers, end-users, and other actors, as previously discussed. In order to be successful in this context, i.e. to create and capture value in the ecosystem, organizations must develop explicit ways and processes to manage change [14, 24]. A specific need for change can be caused by a wide variety of drivers including market changes, legislation, emerging new technologies, and changes in customers' needs. In addition, internal dependencies between the ecosystem members may create a need to change the way the ecosystem reforms itself [25]. Therefore, change management capabilities should be built into the ecosystem's competencies in order that it is in full operational mode when the indicated changes (e.g. due to changed priorities) have to be put in place.

It has been widely seen that there is a favorable impact on the innovation ecosystem related to value and innovation creation, productivity improvement, and overall innovation and business performance [23, 26]. From this perspective, it is essential that ecosystems can systematically develop key enablers of trust-building among the ecosystem members including, e.g. complementarity of obligations regarding the product life cycle, differing perceptions of obligation fulfillment, and balance between value creation, ecosystem objective, and overall mission [29]. Moreover, metaverse ecosystem development involves a wide variety of members, including established companies, universities, the nonprofit sector, and other actors as previously discussed, all of whom share the responsibilities of developing business environments and their related mission and value [24, 25].

From the ecosystem point of view, network operators and solution providers can participate in given opportunities by extending and cost-optimizing their networks using network of networks approaches, enriching NaaS propositions by leveraging Network-as-Code, by providing specialized, dedicated sub-networks for critical industrial needs, or even by directly providing or orchestrating the development of solution platforms or applications. Digital industrial platforms act as both innovation and transaction platforms [30]. First, they allow for the collection and analysis of data from a variety of industrial assets and devices, ranging from tools and machines to vehicles or whole warehouses and factories. These data are usually made available to an ecosystem of third-party firms, which can build complementary solutions such as industrial applications and services. Second, many of the platforms offer marketplaces to facilitate the distribution and use of the created applications by a large market of industrial customers. Thus, digital industrial platforms are an important building block of Industry 4.0 and have influenced the manufacturing industry over the past few years.

Self-organizing networks' self-configuring, optimizing, and healing attributes can allow industry to better plan, deploy, and manage complex 5G network automation. One concrete example of the global ecosystem initiative is the Metaverse Standards Forum, targeted at fostering pragmatic and timely standardization between standards organizations and industry partners. In June 2022, Khronos Group formed and launched the Metaverse Standards Forum. Currently, this nonprofit, open, and member-driven consortium of over 150 companies focuses on royalty-free open standards in areas such as 3D, virtual reality, AR, AI, and parallel programming.

The Metaverse Standards Forum has over 2400 members and is now incorporated as an independent nonprofit industry consortium, bringing companies together in member-driven work groups to find ways forward for the many facets of the metaverse industry. The forum is not, itself, tasked with creating standards. Instead, it aims to gain industry consensus that may lead to specific standards, with 37 principal members providing oversight and governance of the organization.

To realize the virtual–physical fusion and interaction in real-time, the metaverse needs to sense the physical world, including detecting the gestures and actions of a user, the spatial map information, i.e. a 3D map of the indoor or outdoor environment, as well as the positioning of the user in that spatial map. The reconstruction of the physical world includes a large number of computing tasks, such as 3D modeling, object identification and tracking, and virtual scene rendering. Also, the computing tasks need to be completed in real time and sent to the user with low latency. Thus, one of the challenges to support, for example, the XR and metaverse in the 5G-Advanced era and in the future is how to design the network architecture with integrated communication, sensing, and computing functions. With these new capabilities, the following benefits can be realized across the industries. When the industrial metaverse starts to take hold, its scale and complexity will increase exponentially. Enterprises and public sector agencies will be able to create digital twins of entire environments, such as manufacturing sites, power grids, rail yards, or even whole cities, and create, process, and effectively use real-time data to gain unparalleled operational insights in the different industrial environments. Based on this ongoing transformation from industrial 5G toward the industrial metaverse, it is expected that we will see more improvements in safety with remote management, continuous visibility, better training and collaboration opportunities, and more sustainable ways to achieve key business objectives across the industry sectors.

15.3 6G Outlook

The mobile systems landscape has evolved vastly since the first commercial networks. The 1G system, which only supported analog voice service opened the desire for wireless communications that were no longer tied to fixed locations of the traditional networks' devices plugged into home and office walls. Ever since then, the advanced functions and features of 2G, 3G, 4G, and 5G have all been standardized in a parallel fashion.

Figure 15.1 presents the global development of the relative utilization of 2G–5G according to GSMA Intelligence forecasts [31]. It shows that the 2G and 3G networks are losing customer base, and as their sunset advances, 4G will still continue to be the dominant generation well beyond 2025. However, 5G, including its full version based on the Standalone (SA) model, is being deployed on global level and gaining more customer base. GSMA Intelligence predicts that 29% of the customer base will be using 5G by 2025, and in developed markets such as the United States of America, Europe, and Asia, the regional utilization may be close to 50%. By as early as 2030, 5G is expected to account for more than half of all the mobile communications customer base.

While this evolution is advancing, the development of 6G is already underway, although still in an early visioning stage. Until the 3GPP starts the concrete work, among other entities producing technical specifications for 6G, the most

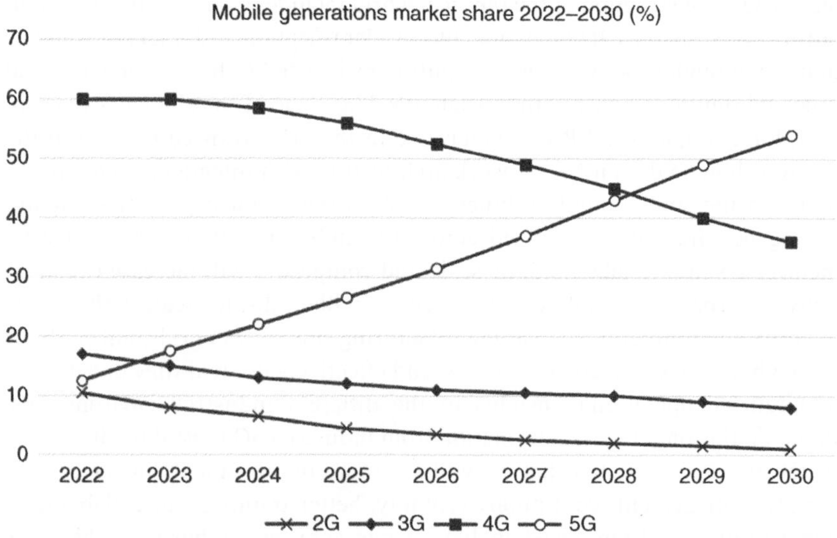

Figure 15.1 Mobile generations' market share, inferred from published information. *Source:* Adapted from GSMA [31].

important is the ITU. It sets the overall requirements for the next generation within the IMT (International Mobile Telecommunications) umbrella of technologies. These requirements will be available for the standard-setting organizations in the IMT-2030 documentation. This is the familiar way that the mobile communications ecosystem has previously complied with the ITU's requirements for 3G, 4G, and 5G systems, in the form of IMT-2000, IMT-Advanced, and IMT 2020 specifications, respectively.[1]

The outcome of the new IMT requirements will be important for the standardization organizations to concretize 6G definitions. The currently available documentation paving the way for the development includes the ITU TS "Architecture Framework" provided in an outlook by the SG13 Study Group in 2020. Meanwhile, the telecom industry is actively envisioning the future and contributing to conceptual testing of solutions beyond 5G (B5G). These initiatives improve the ITU's understanding of the opportunities to use this information, and the technical limitations to use it, in their consideration of realistic IMT requirements for 6G.

Some of the currently available high-level statements regarding 6G indicate that it will connect physical and digital worlds and provide a platform for digital twinning (representing real-world objects in digital forms in near-real time).

Regarding the overall ideas and justifications of 6G, the ITU-T Focus Group Technologies for Network 2030 (FG NET-2030) [18] and Next G Alliance [19], for example, state that 6G requirements will be largely dictated by the use case, which will broaden the previously applied principle of presenting performance characteristics in terms of technical attributes and their values. FG NET-2030 was established by the SG13 ITU-T Study Group in July 2018 and concluded its activity in July 2020. The Next G Alliance is an initiative to advance North American wireless technology leadership over the next decade through private sector-led efforts, encompassing the lifecycle of research and development, manufacturing, standardization, and market readiness.

Furthermore, 6G will integrate physical, biological, and digital worlds along with the evolved radio frequency communication, to facilitate the inclusion of diverse components such as robots, digital twins, AI systems, emotion-driven devices, and brain-machine interfaces [32]. This setup is expected to be capable of enabling a complete cyber-physical-biological communication experience.

---

1 The mapping of ITU IMT variants and corresponding mobile communication generations is for the ecosystem to decide as the ITU does not define the terms 3G–6G. In practice, these globally recognized terms to indicate generations also involve marketing aspects and have room for interpretation, as some concrete systems that claim to represent certain "Gs" might not necessarily formally comply with the respective IMT variant. In this context, the terms IMT-2030 and 6G are assumed to map.

As of yet, ITU-T SG13 has been the key source of information as it provides foundations for the ITU's planning of IMT 6G requirements. This group has discussed 5G capabilities and opportunity areas for the expected 6G use cases, such as holographic communication, that are not straightforward to serve by more limited 5G systems. The ITU SG13 Architecture Framework Specification, published in June 2020, states that 6G will need: the ability to cope with 10–100 times higher data speeds; provide considerably higher system capacity, spectrum efficiency, and radio service coverage; handle faster terminal velocities; and have lower latency, as well as higher capability to interconnect things than previous generations [33, 34].

Nevertheless, during the first half of the 2020, the 6G visions have been under discussion, but clearly 6G will be designed to cope with much more demanding performance requirements than any of the previous generations by the time of its commercial introduction in the beginning of the 2030s. The ITU has already identified candidate architectural models for the more demanding use cases. The ITU's requirements for 6G, referred to as Network 2030 by the SG13 taskforce, are predicted to include simplicity to ensure that the network architectures are manageable. Anticipated NET-2030 requirements also include needs for native programmability and soft re-architecting, backward compatibility, heterogeneous communication, compute, storage, service, and their integration, native slicing, unambiguous naming of network functions and services, intrinsic anonymity and security support for all network operations, resilience, and network determinism.

Along with virtualized network functions, adding services in such a new type of environment may increase complexity with consequent failures, so the solutions under NET-2030 must be kept simple. This means that the architecture must be highly flexible, and Network 2030 must support service and network device decoupling. Furthermore, heterogeneity must be multi-dimensional, native slices should enable efficient use of many types of services, and both users and systems will not access a specific server but rather the hosted content, functions, or services. Moreover, Network 2030 will be a crucial element of security and economic infrastructure (national and global), so it must have adequate means to avoid future cyber-attacks or mitigate their impact through appropriate controls on all sensitive planes of the telecommunication network.

Network 2030 must also be adaptive to accommodate both deterministic and nondeterministic service requirements.

This still rather high-level list of desired capabilities for 6G indicates a need to enhance the current networks considerably in terms of performance, capacity, security, and flexibility. Implications for Industrial IoT and enterprises are twofold as 6G will provide consumers, enterprises, diverse verticals, and other mobile communications users with highly advanced means for connecting services that are challenging, or not even possible, to use fluently in current environments.

References

1 C. Stokel-Walker, “Welcome to the metaverse,” *New Scientist*, vol. 253, no. 3368, pp. 39–43, 2022.
2 J. de la Fuente Prieto, P. Lacasa and R. Martínez-Borda, “Approaching metaverses: Mixed reality interfaces in youth media platforms,” *New Techno Humanities*, vol. 2, no. 2, pp. 136–145, 2022.
3 M. Hernandez-de-Menendez, R. Morales-Menendez, C. Escobar and M. McGovern, “Competencies for Industry 4.0,” *International Journal on Interactive Design and Manufacturing (IJIDeM)*, vol. 14, pp. 1511–1524, 2020.
4 Z. Huang, C. Xiong, H. Ni, D. Wang, Y. Tao and T. Sun, “Standard evolution of 5G-advanced and future mobile network for extended reality and metaverse,” *IEEE Internet of Things Magazine*, vol. 6, no. 1, pp. 20–25, 2023.
5 H. Holma and H. Viswanathan, “In the 6G Era, We Won't Need to Sacrifice Sustainability for the Sake of Performance,” https://www.bell-labs.com/institute/blog/in-the-6g-era-we-wont-need-to-sacrifice-sustainability-for-the-sake-of-performance, 2022.
6 L. Bonati, M. Polese, S. D'Oro, S. Basagni and T. Melodia, “Open, programmable, and virtualized 5G networks: State-of-the-art and the road ahead,” *Computer Networks*, vol. 182, p. 107516, 2020.
7 GSMA, “GSMA Intelligence & Telecommunications,” 2021. [Online].
8 Check Point Research. “Global cyberattacks report.” January 5, 2023. Available: https://blog.checkpoint.com/2023/01/05/38-increase-in-2022-global-cyberattacks/.
9 N. Kshetri, “The economics of the industrial metaverse,” *IT Professional*, vol. 25, no. 1, pp. 84–88, 2023.
10 Z. Zheng, T. Li, B. Li, X. Chai, W. Song, N. Chen, Y. Zhou, Y. Lin and R. Li, “Industrial metaverse: connotation, features, technologies, applications and challenges,” in *Asian Simulation Conference*, (pp. 239–263). Singapore: Springer Nature Singapore, December 9, 2022.
11 ABI-research, “Evaluation of the Enterprise Metaverse Opportunity. Third Quarter 2022,” 2022.
12 MIT Technology Review Insights, “Enabling the next iteration of the internet: The metaverse”. April 11, 2023. Available online: https://www.technologyreview.com/2023/04/11/1069559/enabling-the-next-iteration-of-the-internet-the-metaverse/. [Accessed April 10, 2023].
13 S. Aheleroff, X. Xu, R. Zhong and Y. Lu, “Digital twin as a service (DTaaS) in Industry 4.0: an architecture reference model,” *Advanced Engineering Informatics*, vol. 47, p. 101225, 2021.
14 J. Cao, X. Zhu, S. Sun, Z. Wei, Y. Jiang, J. Wang and V. Lau, “Toward industrial metaverse: age of information, latency and reliability of short-packet transmission in 6G,” *IEEE Wireless Communications*, vol. 30, no. 2, pp. 40–47, 2023.

15 D. Tsolkas and H. Koumaras, “On the development and provisioning of vertical applications in the beyond 5G era,” *IEEE Networking Letters*, vol. 4, no. 1, pp. 43–47, 2022.

16 D. Candal-Ventureira, F. González-Castaño, F. Gil-Castiñeira and P. Fondo-Ferreiro, “Is the edge really necessary for drone computing offloading? An experimental assessment in carrier-grade 5G operator networks,” *Software: Practice and Experience*, vol. 53, no. 3, pp. 579–599, 2023.

17 M. Giess, “CPaaS and SASE: The best defences against IoT threats,” *Network Security*, vol. 9, pp. 9–12, 2021.

18 Z. Lv, S. Xie, Y. Li, M. Hossain and A. El Saddik, “Building the metaverse by digital twins at all scales, state, relation,” *Virtual Reality & Intelligent Hardware*, vol. 4, no. 6, pp. 459–470, 2022.

19 H. Nguyen, R. Trestian, D. To and M. Tatipamula, “Digital twin for 5G and beyond,” *IEEE Communications Magazine*, vol. 59, no. 2, pp. 10–15, 2021.

20 Y. Wang, Z. Su, N. Zhang, R. Xing, D. Liu, T. Luan and X. Shen, “A survey on metaverse: Fundamentals, security, and privacy,” *IEEE Communications Surveys & Tutorials*, vol. 25, no. 1, pp. 319–352, 2022.

21 J. Lee and P. Kundu, “Integrated cyber-physical systems and industrial metaverse for remote manufacturing,” *Manufacturing Letters*, vol. 34, pp. 12–15, 2022.

22 A. Gawer, “Digital platforms’ boundaries: The interplay of firm scope, platform sides, and digital interfaces,” *Long Range Planning*, vol. 54, no. 5, p. 102045, 2021.

23 R. Adner, “Ecosystem as structure: An actionable construct for strategy,” *Journal of management*, vol. 43, no. 1, pp. 39–58, 2017.

24 S. Jung and I. Jeon, “A study on the components of the metaverse ecosystem,” *Journal of Digital Convergence*, vol. 20, no. 2, pp. 163–174, 2022.

25 A. Kar and P. Varsha, “Unravelling the techno-functional building blocks of metaverse ecosystems – A review and research agenda,” *International Journal of Information Management Data Insights*, p. 100176, 2023.

26 L. Linde, D. Sjödin, V. Parida and J. Wincent, “Dynamic capabilities for ecosystem orchestration A capability-based framework for smart city innovation initiatives,” *Technological Forecasting and Social Change*, vol. 166, p. 120614, 2021.

27 J. Pellikka and T. Ali-Vehmas, “Managing innovation ecosystems to create and capture value in ICT industries,” *Technology Innovation Management Review*, vol. 6, no. 10, pp. 17–24, 2021.

28 K. Möller, S. Nenonen and K. Storbacka, “Networks, ecosystems, fields, market systems? Making sense of the business environment,” *Industrial Marketing Management*, vol. 90, pp. 380–399, 2020.

29 N. Foss, J. Schmidt and D. Teece, “Ecosystem leadership as a dynamic capability,” *Long Range Planning*, vol. 56, no. 1, p. 102270, 2022.

30 M. Cusumano, "The evolution of research on industry platforms," *Academy of Management Discoveries*, vol. 8, no. 1, pp. 7–14, 2022.

31 GSMA, "The Mobile Economy Report 2023," GSMA, 2023. [Online]. Available: https://www.gsma.com/mobileeconomy/wp-content/uploads/2023/03/270223-The-Mobile-Economy-2023.pdf. [Accessed 27 April 2023].

32 J. R. Bhat, "6G ecosystem: current status and future," *IEEE Access*, vol. 9, no. 2021, p. 43134–43167, 26 January 2021.

33 ITU-T FG-NET2030 Focus Group for Network 2030, "Network 2030 Architecture Framework," *Technical Specification*, June 2020. Available: https://www.itu.int/en/ITU-T/focusgroups/net2030/Documents/Network_2030_Architecture-framework.pdf. [Accessed 27 October 2023].

34 J. Penttinen, "On 6G Visions and Requirements," *Journal of ICT Standardization 2021*, vol. 9, no. 3. Published 22 December 2022 [Online]. Available: https://journals.riverpublishers.com/index.php/JICTS/article/view/6903. [Accessed 27 October 2023].

Index

5G Innovations for Industry Transformation: Data-Driven Use Cases, First Edition.
Jari Collin, Jarkko Pellikka, and Jyrki T.J. Penttinen.

Published 2024 by John Wiley & Sons, Inc.

Printed and bound by CPI Group (UK) Ltd, Croydon, CR0 4YY

27/10/2024

14580669-0003